Methods for Solving Systems of Nonlinear Equations

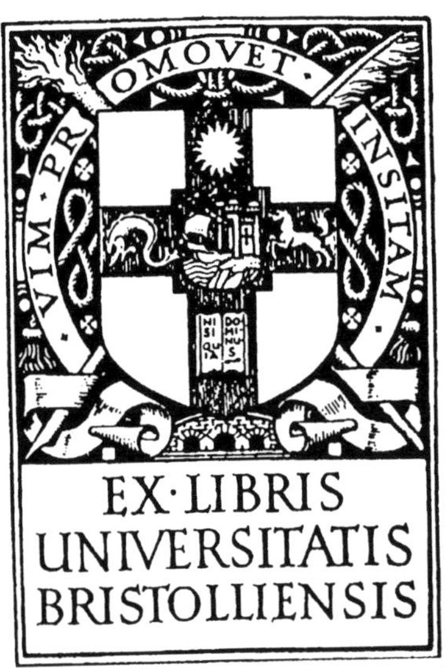

Werner C. Rheinboldt
University of Pittsburgh
Pittsburgh, Pennsylvania

Methods for Solving Systems of Nonlinear Equations

Second Edition

siam.
SOCIETY FOR INDUSTRIAL AND APPLIED MATHEMATICS

PHILADELPHIA

10 9 8 7 6 5 4 3 2 1

Library of Congress Cataloging-in-Publication Data

Rheinboldt, Werner C.
Methods for solving systems of nonlinear equations / Werner C. Rheinboldt. -- 2nd ed.
p. cm. -- (CBMS-NSF regional conference series in applied mathematics ; 70)
Includes bibliographical references and index.
ISBN 0-89871-415-X (pbk.)
1. Equations, Simultaneous--Data processing. 2. Nonlinear theories--Data processing. 3. Numerical analysis--Data processing. I. Title. II. Series.
QA214.R44 1998
512.9'4--dc21 98-27321

Contents

Preface to the Second Edition

After SIAM's editors asked me to prepare a second edition of this monograph, it became clear to me that the book needed to become more self-contained and that the inclusion of some proofs would be essential. Accordingly, this second edition has been extensively reworked and expanded. Some of the material has been rearranged to accommodate additional topics and to deemphasize others. But, throughout, the aim has remained to highlight the ideas behind the algorithms as well as their theoretical foundations and properties rather than to concentrate on the computational details. This reflects my view that—due to the predominantly analytic nature of this area—it is the structure of the theoretical development that provides a major part of the required insight.

The literature on the computational solution of systems of nonlinear equations has grown considerably since the first edition, and it would have been impossible to accommodate here the numerous advances and new results. Thus many decisions had to be made to forgo adding material that would simply exceed the framework or to make changes that would break the format of the existing presentation. This is especially reflected in the chapter on minimization methods, where the feasible number of changes and additions falls far short of the range of new material. It also precluded the addition of entirely new topics such as methods for large systems of equations, or numerical examples illustrating computational experience. Nevertheless, there are additions to almost all chapters. This includes an introduction to the theory of inexact Newton methods, a basic theory of continuation methods in the setting of differentiable manifolds, and an expanded discussion of minimization methods.

It is hoped that this second edition indeed provides the needed update of this introduction to the numerical analysis of an important computational area.

Preface to the First Edition

This monograph contains a written and somewhat expanded version of the invited lectures I presented at the NSF-CBMS Regional Conference on "The Numerical Solution of Nonlinear Algebraic Systems," which was held July 10–14, 1972, at the University of Pittsburgh.

The aim of the conference was to acquaint applied scientists with some of the principal results and recent developments in the computational solution of nonlinear equations in several variables. Accordingly, these lecture notes represent in part a survey of fundamental aspects of the field as well as of typical applications and in part a discussion of several important topics of current research interest. As in the earlier volumes of this monograph series, the stress has been placed on general concepts, results, and applications, with specific references given to detailed proofs elsewhere.

It is a sincere pleasure to acknowledge the support of the National Science Foundation for this Regional Conference under Grant GJ-33612 to the Conference Board of the Mathematical Sciences and the University of Pittsburgh, as well as for the part of my own research that is reflected in these lecture notes, under grant GJ-35568X. I would also like to express my special thanks to the Conference Committee under the chairmanship of Professor Charles A. Hall for the excellent planning and superb handling that made this conference such a success. And my warm thanks also go to Mrs. Dawn Shifflett for her untiring and careful help in the preparation of the manuscript.

CHAPTER 1

Introduction

As an introduction to our topic we begin this chapter with a brief overview of various questions that arise in connection with the computational solution of nonlinear systems of equations. Then, for ease of reference, we collect some notational conventions and background results from linear algebra and analysis. In this and all subsequent chapters we have frequent occasion to refer to the monograph by Ortega and Rheinboldt [192]. In order to avoid repetitious citations that book shall be identified simply by [OR].

1.1 Problem Overview

Let $F : \mathrm{R}^n \mapsto \mathrm{R}^n$ be a nonlinear mapping from the n-dimensional real linear space R^n into itself. Our interest is in methods for the computation of solutions of the system of n equations in n variables

$$Fx = 0. \tag{1.1}$$

This involves at least two overlapping problem areas, namely, (a) the analysis of the solvability properties of the system, and (b) the development and study of suitable numerical methods.

Clearly, meaningful results for both of these problems depend critically on the properties of the function F. Two special cases of (1.1) are much better understood than most others, namely, the n-dimensional linear systems, and the one-dimensional (i.e., scalar) nonlinear equations. We refer, e.g., to the texts of Golub and Van Loan [114], Stewart [256], and Young [284] for the first case and to those of Traub [265], Householder [133], and Heitzinger, Troch, and Valentin [124] for the second one. We shall not dwell on results that are specific only to these special cases but focus instead on methods for general nonlinear systems (1.1).

Under the general heading of a solvability analysis of the system (1.1) we have to consider at least three questions:

(a) Do solutions exist in a specified subset of the domain of F?

(b) How many solutions are there in such a set?

(c) How do the solutions vary under small changes to the problem?

Already in the case of scalar equations, simple examples show that (1.1) may have either no solution, or any finite or even infinite number of them. In fact, one can prove that for any closed set $S \subset \mathrm{R}^1$ there exists a C^∞ map from R^1 into itself which has S as its zero set. This is a special case of a more general result of Whitney [278]. It is also easy to give examples where the solutions exhibit rapid changes under continuous variation of F. Thus, in general, some solvability analysis is very much needed. This represents a topic of nonlinear functional analysis. While we cannot enter into a discussion of these solvability questions, it may be useful to mention at least briefly some of the principal approaches that are applicable in the finite-dimensional case (see also [OR] for a more extensive introduction).

A conceptually simple yet powerful technique is the transformation of (1.1) into a minimization problem for some nonlinear functional (see also Chapter 8). This corresponds to the variational approach in the theory of differential equations. Other existence results for (1.1) are based on arguments deriving from the contraction principle and its many generalizations. Besides the contraction-mapping theorem itself (see Theorem 4.1), a typical local result is, of course, the inverse-function theorem. Somewhat deeper are various topological theories used in nonlinear functional analysis, ranging from classical degree theory to modern results of differential topology and global analysis (see, e.g., Aubin [13], Aubin and Ekeland [14], and Berger [22]). Here, typical results include the Brouwer fixed-point theorem (see Theorem 10.2) and the domain-invariance theorem, which states that when $E \subset \mathrm{R}^n$ is open and $F : E \mapsto \mathrm{R}^n$ is one-to-one and continuous then the image set $F(E)$ is open and F is a homeomorphism. Further examples are the various fixed-point theorems for set-valued mappings.

Provided the equation (1.1) has solutions, our principal concern will be to introduce some of the methods for approximating them. These methods might be distinguished by their aims, such as

(a) Localization methods for constructing sets containing solutions;
(b) Methods for approximating one solution;
(c) Methods for finding all solutions.

Our discussion here will be restricted to (b). Methods that meet aim (c) are essentially available only for minimization problems (see section 10.3 for some references). Most of the methods under (a) are based on the use of interval computations. In recent years this has become an active area of research, and the resulting computational methods are proving themselves to be very effective (see, e.g., the recent monograph of Kearfott [141]).

Except for special cases, such as linear systems, direct methods for solving (1.1) are generally not feasible and attention therefore must focus on iterative processes. A broad classification of these processes may be based on the principles used to generate them:

(a) Linearization methods (see Chapters 4 and 5);
(b) Reduction to simpler nonlinear equations (see section 6.2);
(c) Combination of processes (see Chapter 6);
(d) Continuation methods (see Chapter 7);
(e) Minimization methods (see Chapter 8).

This classification is useful even though there is some overlap, and, as might be expected, there are methods which do not fit conveniently into any one of the classes, notably those designed for specific types of equations.

A discussion of iterative processes should respond, at least in part, to the following four general questions.

(a) Under what conditions are the iterates well defined?

(b) When do the iterates converge?

(c) In the case of convergence, is the limit a solution of (1.1)?

(d) How economical is the process?

In most instances only partial answers can be given to these questions. In particular, the exact sets of all initial data for which a process is well defined or for which it converges are very rarely computable.

Outside the area of minimization methods there are three overlapping types of convergence results. The *local convergence theorems* assume the existence of a particular solution x^* and then assert the existence of a neighborhood $\mathcal{U}$ of x^* such that, when started anywhere in $\mathcal{U}$, the iterates are well defined and converge to x^*. The *semilocal theorems* do not require knowledge of the existence of a solution; instead they state that, for initial data satisfying certain—usually stringent—conditions, there exists a (typically nearby) solution to which the process will converge. Finally, the most desirable convergence results, the *global theorems*, assert that starting (essentially) anywhere in the domain of the operator, convergence to some solution is assured—although it may not be the one that had been sought. We concentrate here on the first and last of these types of results and, for space reasons, forgo to address any of the semilocal theorems.

Many examples show that the limit of an iteration may exist but does not solve the given system. Thus any convergence result will be acceptable only if it asserts that the resulting limits are indeed solutions of (1.1).

In connection with question (d) about the economics or computational complexity of the process we may distinguish several interrelated subquestions:

(a) What is the "cost" of one step of the iterative process?

(b) How "fast" is the sequence expected to converge?

(c) How "sensitive" is the process to changes of the mapping, the initial data, or the dimension of the space?

Any answers to these questions require quantitative measures for the terms "cost", "fast", and "sensitive", and these, in turn, demand a precise specification of the data used by the process. In recent years, an extensive theory of *information-based complexity* has been developed which allows for the characterization of optimal members of certain classes of processes (see, e.g., the monographs by Traub and Woźniakowski [267], and Traub, Wasilkowski, and Woźniakowski [266]). We restrict ourselves here to the less sweeping but nevertheless informative results on the *convergence rates* of specific processes based on certain asymptotic measures of convergence (see section 3.2). Question (c) appears to have the fewest general answers, although, for certain types of problems, many algorithms certainly show extreme sensitivity to changes of the operator or the initial data. This may often be explained by the presence of certain singularities and, in fact, the study of such singularities is an important field of study of its own.

1.2 Notation and Background

The presentation assumes a basic background in linear algebra and analysis. For ease of reference we collect here some notational conventions and frequently used background results.

The real n-dimensional linear space of column vectors x with components $x_1, x_2, \ldots, x_n$ is denoted by R^n, and C^n is the corresponding space of complex column vectors. These spaces are endowed with their natural topology which defines such concepts as open and closed sets; the interior, int (S), of a set S; neighborhoods of a point; limits of sequences of vectors; etc. The canonical (natural) basis of R^n will be denoted by $e^1, \ldots, e^n$. For $x \in \mathrm{R}^n$ or $x \in \mathrm{C}^n$, $x^\top$ or x^H is the transpose or Hermitian transpose of x, respectively.

A real $m \times n$ matrix $A = (a_{ij})$ defines a linear mapping from R^n to R^m and we write $A \in L(\mathrm{R}^n, \mathrm{R}^m)$ to denote either the matrix or the linear operator, as context dictates. Similarly, $L(\mathrm{C}^n, \mathrm{C}^m)$ is the linear space of complex $m \times n$ matrices. In the case $n = m$, the abbreviations $L(\mathrm{R}^n)$ and $L(\mathrm{C}^n)$ will be used. Clearly, as R^n can be imbedded in C^n, we may similarly consider $L(\mathrm{R}^n, \mathrm{R}^m)$ to be naturally imbedded in $L(\mathrm{C}^n, \mathrm{C}^m)$. The space $L(\mathrm{R}^n, \mathrm{R}^1)$ is isomorphic with the space of all row vectors. For any $A \in L(\mathrm{C}^n, \mathrm{C}^m)$ the nullspace and range space are denoted by ker A and rge A, respectively, and rank A is the rank of A; that is, the dimension of rge A. By $A^\top \in L(\mathrm{R}^m, \mathrm{R}^n)$ or $A^H \in L(\mathrm{C}^m, \mathrm{C}^n)$ we mean the transpose or Hermitian transpose, respectively. The linear subspaces of $L(\mathrm{R}^n)$ or $L(\mathrm{C}^n)$ consisting of all symmetric or Hermitian matrices are denoted by $L_S(\mathrm{R}^n)$ and $L_H(\mathrm{C}^n)$, respectively.

A linear operator $A \in L(\mathrm{R}^n)$ is invertible (or nonsingular) if it is one-to-one, in which case its inverse is written as A^{-1}. Of the various perturbation results for invertible matrices we quote the Sherman–Morrison–Woodbury formula. For invertible $A \in L(\mathrm{R}^n)$ and $U, V \in L(\mathrm{R}^m, \mathrm{R}^n)$, $m \leq n$, the matrix $A + UV^\top$ is invertible exactly if $I_n + V^\top A^{-1} U$ is invertible and, in that case

$$(A + UV^\top)^{-1} = A^{-1} - A^{-1} U (I_n + V^\top A^{-1} U)^{-1} V^\top A^{-1}.$$

An eigenvalue of $A \in L(\mathrm{C}^n)$ is any $\lambda \in \mathrm{C}^1$ such that $Ax = \lambda x$ for some nonzero $x \in \mathrm{C}^n$, called an eigenvector of A corresponding to λ. The set of all eigenvalues is the spectrum $\sigma(A) \subset \mathrm{C}^1$ of A. It has at least one and at most n distinct members and $\rho(A) = \max\{|\lambda| : \lambda \in \sigma(A)\}$ is the spectral radius of A.

Any otherwise unspecified norm on R^n will be denoted by $\|\cdot\|$. Frequently used norms include the ℓ_p-norms and their limiting case, the ℓ_∞-norm; that is,

$$\|x\|_p = \left[\sum_{k=1}^{n} |x_k|^p \right]^{\frac{1}{p}}, \quad 1 \leq p < \infty, \quad \|x\|_\infty = \sup_{k=1,\ldots,n} |x_k|, \quad \forall x \in \mathrm{R}^n.$$

The ℓ_2-norm is induced on R^n or C^n by the natural Euclidean inner product $\langle x, y \rangle = x^\top y$ or $\langle x, y \rangle = x^H y$, respectively. Under this inner product such concepts as orthogonality of vectors or linear subspaces, and the orthogonal projections are defined.

Any two norms $\|\cdot\|_a$ and $\|\cdot\|_b$ on R^n are equivalent; that is, they induce the same topology, or, equivalently, there exist for them constants $c_2 \geq c_1 > 0$ such that

$$c_1\|x\|_b \leq \|x\|_a \leq c_2\|x\|_b, \quad \forall x \in \mathrm{R}^n. \tag{1.2}$$

A matrix norm $\|\cdot\|$ on $L(\mathrm{R}^n, \mathrm{R}^m)$ is consistent with given vector norms on R^n and R^m if

$$\|Ax\| \leq \|A\|\,\|x\|, \quad \forall x \in \mathrm{R}^n.$$

Analogously, three matrix norms on $L(\mathrm{R}^n, \mathrm{R}^p)$, $L(\mathrm{R}^p, \mathrm{R}^m)$, $L(\mathrm{R}^n, \mathrm{R}^m)$, respectively, are consistent if

$$\|AB\| \leq \|A\|\,\|B\|, \quad \forall A \in L(\mathrm{R}^n, \mathrm{R}^p),\ B \in L(\mathrm{R}^p, \mathrm{R}^m).$$

Any vector norm on R^n and R^m induces a matrix norm

$$\|A\| = \sup_{x \neq 0} \frac{\|Ax\|}{\|x\|}, \quad A \in L(\mathrm{R}^n, \mathrm{R}^m),$$

which is consistent with the two vector norms. For $m = 1$ the induced norm on $L(\mathrm{R}^n, \mathrm{R}^1)$ is the dual norm of $\|\cdot\|$. For any matrix norm on $L(\mathrm{R}^n)$ which is consistent with itself (multiplicative norm) there exists a vector norm on R^n which is consistent with it. Any inner product on $L(\mathrm{R}^n)$ defines an associated norm on $L(\mathrm{R}^n)$. An example is the Frobenius inner product

$$\langle A, B \rangle_F = \mathrm{trace}(A^\top B), \quad \forall A, B \in L(\mathrm{R}^n, \mathrm{R}^m), \tag{1.3}$$

which has as associated norm the Frobenius norm

$$\|A\|_F = [\mathrm{trace}(A^\top A)]^{\frac{1}{2}} = \left[\sum_{i=1}^{n}\sum_{k=1}^{m} |a_{i,k}|^2\right]^{\frac{1}{2}}, \quad \forall A \in L(\mathrm{R}^n, \mathrm{R}^m).$$

Here we used the trace operator $A \in L(\mathrm{R}^n) \mapsto \mathrm{trace}\,(A) = a_{11} + \cdots + a_{nn}$ which is linear in its argument and also satisfies $\mathrm{trace}\,(AB) = \mathrm{trace}\,(BA)$, $\forall A, B \in L(\mathrm{R}^n)$. The Frobenius norm is consistent with itself and consistent with the ℓ_2-norms on R^n and R^m, but it is not an induced norm.

A frequently used result using matrix norms is the "perturbation lemma": If $A \in L(\mathrm{R}^n)$ is invertible with $\|A^{-1}\| \leq \alpha$ for some multiplicative norm, then any $B \in L(\mathrm{R}^n)$ such that $\|A - B\| \leq \beta$ and $\alpha\beta < 1$ is invertible and $\|B^{-1}\| \leq \alpha/(1 - \alpha\beta)$.

A mapping

$$F : E \subset \mathrm{R}^n \mapsto \mathrm{R}^m,\ E \text{ open},\ Fx = (f_1(x), \ldots, f_m(x))^\top,\ \forall x \in E \tag{1.4}$$

is of class C^r, $r \geq 0$ (or C^r for short) on E if, for $i = 1, \ldots, m$, the partial derivatives of order at least r of each component functional f_i exist and, together with f_i, are continuous on E. The mapping F is of class C^∞ if it is of class C^r for each positive r.

These differentiability definitions are coordinate dependent. We note only the following coordinate-independent definitions of first derivatives.

A mapping (1.4) is *Gateaux-differentiable* (*G-differentiable*) at $x \in E$ if there exists a linear operator $A \in L(\mathrm{R}^n, \mathrm{R}^m)$ such that

$$\lim_{t \to 0} \frac{1}{t} \|F(x + th) - Fx - tAh\| = 0, \quad \forall\, h \in \mathrm{R}^n.$$

The limit is independent of the norm on R^m and the operator A is unique; it is called the *G-derivative* of F at x and denoted by $DF(x)$. With the bases used in (1.4) the matrix representation of $DF(x)$ is the Jacobian matrix

$$DF(x) = \begin{pmatrix} \frac{\partial}{\partial x_1} f_1(x) & \cdots & \frac{\partial}{\partial x_n} f_1(x) \\ \cdots & \cdots & \cdots \\ \frac{\partial}{\partial x_1} f_m(x) & \cdots & \frac{\partial}{\partial x_n} f_m(x) \end{pmatrix}.$$

The mapping (1.4) is *Frechet-differentiable* (*F-differentiable*) at $x \in E$ if there exists a linear operator $A \in L(\mathrm{R}^n, \mathrm{R}^m)$ such that

$$\lim_{h \to 0} \frac{1}{\|h\|} \|F(x + th) - Fx - Ah\| = 0.$$

Clearly, F is G-differentiable at x whenever it is F-differentiable at that point and hence A is unique and equal to $DF(x)$.

Consider the three statements (i) F is C^1 on a neighborhood of x, (ii) F is F-differentiable at x, (iii) F is G-differentiable at x. Then it follows that (i) $\Rightarrow$ (ii) $\Rightarrow$ (iii) with the same $A = DF(x)$. But, generally, the converse implications do not hold. Moreover, (ii) implies that F is continuous at x while (iii) only provides that F is hemicontinuous at x, in other words, that $\lim_{t \to 0} F(x + th) = Fx$ for any fixed $h \in \mathrm{R}^n$.

Higher derivatives are defined recursively. For $DF : \mathrm{R}^n \mapsto L(\mathrm{R}^n, \mathrm{R}^m)$ we have $D^2F(x) \in L(\mathrm{R}^n, L(\mathrm{R}^n, \mathrm{R}^m))$. Since $L(\mathrm{R}^n, L(\mathrm{R}^n, \mathrm{R}^m))$ is isomorphic with the space $L^2(\mathrm{R}^n, \mathrm{R}^m)$ of all bilinear maps from $\mathrm{R}^n \times \mathrm{R}^n$ to R^m we always consider $D^2F(x)$ as a member of $L^2(\mathrm{R}^n, \mathrm{R}^m)$. Correspondingly, $D^pF(x)$ will be considered as a member of the space $L^p(\mathrm{R}^n, \mathrm{R}^m)$ of all p-linear maps from $\mathrm{R}^n \times \cdots \times \mathrm{R}^n$ to R^m. For a functional f on R^n, the matrix representation of the second derivative is $D^2f(x)(h, k) = k^\top H_f(x)h$, $h, k \in \mathrm{R}^n$, where

$$H_f(x) = \begin{pmatrix} \frac{\partial^2}{\partial x_1 \partial x_1} f(x) & \cdots & \frac{\partial^2}{\partial x_n \partial x_1} f(x) \\ \cdots & \cdots & \cdots \\ \frac{\partial^2}{\partial x_1 \partial x_n} f(x) & \cdots & \frac{\partial^2}{\partial x_n \partial x_n} f(x) \end{pmatrix}$$

is the Hessian matrix of f at x.

If F is G-differentiable on a convex set $E_c \subset \mathrm{R}^n$ then

$$\|Fy - Fx\| \le \sup_{0 \le t \le 1} \|DF(x + t(y - x))\| \, \|y - x\|, \quad \forall\, x, y \in E_c.$$

Hence, if $\|DF(x)\| \leq \gamma < \infty$ for $x \in E_c$, then F is Lipschitz-continuous on that set. On the other hand, if F is C^1 on an open convex set E_c then we have the integral mean value theorem

$$Fy - Fx = \int_0^1 DF(x + t(y - x))(y - x)dt, \quad \forall\, x, y \in E_c.$$

Moreover, when F is C^2 then

$$Fy - Fx - DF(x)(y - x) = \int_0^1 (1 - t)DF(x + t(y - x))(y - x)(y - x)dt$$

holds for all $x, y \in E_c$.

CHAPTER 2

Model Problems

Systems of finitely many nonlinear equations in several real variables arise in connection with numerous scientific and technical problems and, correspondingly, they differ widely in form and properties. This chapter introduces some typical examples of such systems without attempting to be exhaustive or to enter into details of the underlying problem areas. For further nonlinear model problems see [OR] and the collection by Moré [179].

2.1 Discretization of Operator Equations

The computational solution of infinite-dimensional operator equations generally requires the construction of finite-dimensional approximations. It is reasonable to expect that these discretizations inherit any nonlinearities existent in the original operators. Here is one of the principal sources for nonlinear equations in finitely many variables.

As a simple example, let Ω be a bounded domain in R^d with (Lipschitz) boundary $\partial\Omega$, and consider the boundary-value problem

$$(2.1) \qquad -\Delta u = \lambda e^u \text{ in } \Omega, \quad u = 0 \text{ on } \partial\Omega, \qquad \Delta = \sum_{i=1}^{d} \frac{\partial^2}{\partial \xi_i^2},$$

where λ is a given parameter. This is usually called Bratu's problem and arises in the simplification of models of nonlinear diffusion processes, see, e.g., Aris [9] for some connections with combustion problems.

A simple approach to the numerical solution of (2.1) is based on the use of the classical five-point difference approximation for Δu. For ease of notation, let $\Omega = [0,1] \times [0,1] \subset \mathrm{R}^2$ and consider a uniform mesh

$$(2.2) \qquad p_{i,j} = (ih, jh), \quad h = \frac{1}{m+1}, \quad i,j = 0,1,\ldots,m+1.$$

The Laplacian Δu at the mesh-point p_{ij} may be approximated by the difference quotient

$$(2.3) \qquad \frac{1}{h^2}[u(p_{i+1,j}) + u(p_{i-1,j}) + u(p_{i,j+1}) + u(p_{i,j-1}) - 4u(p_{i,j})].$$

In other words, approximations

$$x_1 \doteq u(p_{11}), \ldots, x_m \doteq u(p_{m1}), x_{m+1} \doteq u(p_{12}), \ldots, x_n \doteq u(p_{mm}), \; n = m^2,$$

of the values of the desired solution u at the mesh-points (2.2) may be defined as the components of the solution vector $x = (x_1, \ldots, x_n)^\top \in \mathrm{R}^n$ of the nonlinear system of equations

$$Fx \equiv Ax - \Phi x = 0, \tag{2.4}$$

resulting from (2.1) by replacing Δu with (2.3). Here, the matrix $A \in L(\mathrm{R}^n)$ has the block form

$$A = \begin{pmatrix} B & -I & \ldots & 0 \\ -I & B & \ldots & 0 \\ \vdots & \vdots & \ddots & \vdots \\ 0 & 0 & \ldots & B \end{pmatrix}, \quad B = \begin{pmatrix} 4 & -1 & \ldots & 0 \\ -1 & 4 & \ldots & 0 \\ \vdots & \vdots & \ddots & \vdots \\ 0 & 0 & \ldots & 4 \end{pmatrix} \in L(\mathrm{R}^m), \tag{2.5}$$

and the nonlinear mapping Φ is

$$\Phi : \mathrm{R}^n \mapsto \mathrm{R}^n, \quad \varphi_i(x) = h^2 \lambda e^{x_i}, \quad i = 1, \ldots n. \tag{2.6}$$

The system (2.4), (2.5), (2.6) is an example of a particularly simple class of nonlinear equations. Generally, we introduce the following concept.

DEFINITION 2.1. *The pattern matrix of* $F : E \subset \mathrm{R}^n \mapsto \mathrm{R}^n$ *is the matrix* $T_F = (\tau_{i,j}) \in L(\mathrm{R}^n)$ *with* $\tau_{i,j} = 0$ *when the* i^{th} *component function* f_i *of* F *is independent of* x_j *in all of* E, *and* $\tau_{i,j} = 1$ *otherwise.*

It is natural to call a mapping diagonal if its pattern matrix is diagonal. An example of this is the mapping (2.6). An operator F is called *almost linear* if it is the sum of a linear operator and a diagonal nonlinear mapping. An example is, of course, our system (2.4), (2.5), (2.6). Such systems are amenable to many numerical approaches and are valuable as computational test problems.

The situation changes if in (2.1) the exponential on the right side is replaced by a function that depends not only on u but also on certain first derivatives $\partial u / \partial \xi_j$. In that case, the use of (2.3) and, say, of symmetric difference quotients for the first derivative terms, again leads to a system of the form (2.4); but now the ith component φ_i of the mapping Φ depends also on variables other than x_i, and hence Φ is no longer diagonal. Here, in fact, the pattern matrix will typically be tridiagonal.

When the original operator equation has a more complicated form than (2.1) the same will be true for its finite-dimensional analogues. For example, on a uniform mesh

$$\xi_i = ih, \quad h = \frac{1}{n+1}, \quad i = 0, \ldots, n+1, \tag{2.7}$$

the quasi-linear problem

$$-\frac{d}{d\xi}\left(p(u)\frac{du}{d\xi}\right) = f(\xi, u), \qquad 0 < \xi < 1, \quad u(0) = u(1) = 0$$

may be approximated by the difference equations

$$(2.8) \qquad -\frac{1}{h}\left[p(x_{i+1})\frac{x_{i+1}-x_i}{h} - p(x_i)\frac{x_i - x_{i-1}}{h}\right] = f(\xi_i, x_i), \; i = 1, \ldots, n,$$

in which $x_0 = x_{n+1} = 0$. Evidently, (2.8) has the form

$$(2.9) \qquad Fx \equiv A(x)x - \Phi x = 0,$$

where the matrix $A(x) = (a_{ij}(x)) \in L(\mathrm{R}^n)$ has the elements

$$a_{ij}(x) = \frac{1}{h^2}\begin{cases} -p(x_i) & \text{if } j = i-1 \\ p(x_i) + p(x_{i+1}) & \text{if } j = i \\ -p(x_{i+1}) & \text{if } j = i+1 \\ 0 & \text{otherwise} \end{cases} \qquad i, j = 1, \ldots, n,$$

and the mapping Φ has the components $\varphi_i(x) = f(ih, x_i)$, $i = 1, \ldots, n$. In line with the terminology for differential operators, (2.9) is usually called a *quasi-linear* system of equations (see, e.g., Meyer [169]).

Finite-difference methods involve a discretization of both the domain and the operator and provide values of the approximate solution only at a finite set of mesh-points. Moreover, they typically require the solution to be sufficiently smooth; see, e.g., the books by Ames [8], Forsythe and Wasow [101], Marchuk and Shaidurov [162], Marchuk [161], Mitchell and Griffiths [172], or Törnig, Gipser, and Kaspar [264].

Often preferable, certainly in cases of domains of complicated shape or of solutions with lower smoothness properties, are finite-element methods for which there exists a very extensive literature. They are based on a weak or variational formulation of the given infinite-dimensional problem and lead to discrete analogues by a discretization of the solution space.

For example, a weak form of Bratu's problem (2.1) requires finding a function $u \in H_0^1(\Omega)$ such that

$$(2.10) \qquad \int_\Omega \nabla u \cdot \nabla v \, d\xi = \lambda \int_\Omega e^u v \, d\xi, \quad \forall v \in H_0^1(\Omega),$$

(see, e.g., Glowinski [110]). Here, the standard notation

$$d\xi = d\xi_1 \ldots d\xi_d, \quad \nabla = (\frac{\partial}{\partial \xi_1}, \ldots, \frac{\partial}{\partial \xi_d}), \quad \nabla u \cdot \nabla v = \sum_{i=1}^{d} \frac{\partial u}{\partial \xi_i}\frac{\partial v}{\partial \xi_i},$$

was used, and $H_0^1(\Omega)$ is the Sobolev space consisting of all $w \in L^2(\Omega)$ such that $w = 0$ on $\partial\Omega$ and that the first derivatives in the distributional sense exist and satisfy $\frac{\partial w}{\partial \xi_i} \in L^2(\Omega)$, $i = 1, \ldots, d$. Now, for the discretization an appropriate subspace $X_n \subset H_0^1(\Omega)$ of finite dimension n is constructed and (2.10) is replaced by the approximate problem of finding a function $w \in X_n$ such that

$$(2.11) \qquad \int_\Omega \nabla w \cdot \nabla v \, d\xi = \lambda \int_\Omega e^u \; v \, d\xi, \quad \forall v \in X_n.$$

For finite-element methods the space X_n typically consists of piecewise polynomials on a "triangulation" of the domain. As noted, the literature is vast and we refer, e.g., to the books by Axelson and Barker [15], Ciarlet [50], Schwarz [237], Szabo and Babuska [258], and Zienkiewicz [286]. Here we include only the simple example of Bratu's problem (2.10) with $d = 1$ on the interval $\Omega = [0, 1] \subset \mathrm{R}^1$. On the uniform mesh (2.7) define the "hat-functions"

$$w_i(\xi) = \frac{1}{h}\begin{cases} \xi - \xi_{i-1} & \text{for } \xi_{i-1} \le \xi \le \xi_i \\ \xi_{i+1} - \xi & \text{for } \xi_i \le \xi \le \xi_{i+1} \\ 0 & \text{otherwise} \end{cases} \qquad i, j = 1, \ldots, n, \tag{2.12}$$

and with them the subspace

$$X_n = \left\{ w \in H_0^1(\Omega) : \; w = \sum_{i=1}^{n} x_i w_i, \; x \in \mathrm{R}^n \right\}.$$

Then, with $w \in X_n$ and $v = w_k$ substituted in (2.11), a simple calculation gives, for $k = 1, \ldots, n$,

$$-x_{k-1} + 2x_k - x_{k+1} = h^2 \int_0^1 t e^{t x_k} [e^{(1-t)x_{k-1}} + e^{(1-t)x_{k+1}}] dt, \; x_0 = x_{n+1} = 0.$$

This is a nonlinear system in $x = (x_1, \ldots, x_n)^\top$ of the general form (2.4) but, of course, the nonlinear mapping Φ is no longer diagonal.

A different approach to the numerical solution of two-point boundary-value problems is the shooting method. As an example, consider the system of ordinary differential equations

$$\begin{cases} u' = f(s, u), \quad 0 < s < 1, \quad u(s) \in \mathrm{R}^n, \\ Au(0) + Bu(1) = a, \quad a \in \mathrm{R}^n, \\ A, B \in L(\mathrm{R}^n), \quad \text{rank } (A, B) = n. \end{cases} \tag{2.13}$$

Assume that for any $x \in \mathrm{R}^n$ the associated initial-value problem

$$z' = f(s, z), \quad z(0) = x, \tag{2.14}$$

has a unique solution $z = z(s, x)$ for $s \in [0, 1]$. Then the mapping $F : \mathrm{R}^n \mapsto \mathrm{R}^n$, $Fx = Ax + Bz(1, x) - a$ is well defined and $u = z(s, x^*)$ is a solution of (2.13) whenever x^* solves $Fx = 0$.

An evaluation of F requires the solution of the initial value problem (2.14). In practice, this involves the use of some step-by-step procedure which produces an approximation $\hat{z}(s, x)$ of $z(s, x)$. Therefore, instead of the system $Fx = 0$, we solve some related system $\hat{F}x \equiv Ax + B\hat{z}(1, x) - a = 0$ in which $\hat{F}$ is defined—albeit in a complicated manner—by the integration procedure.

The basic shooting method is prone to numerical instabilities caused by a possible growth of the solutions of (2.14). In order to reduce this growth, variants of the method, such as the parallel-shooting method, have been designed. Let

$$0 = s_0 < s_1 < \ldots < s_{m-1} < s_m = 1, \quad h_j = s_j - s_{j-1}, \quad j = 1, \ldots, m,$$

be a given mesh, and consider the system of initial-value problems

$$\frac{dz^j}{dt} = h_j f(s_{j-1} + th_j, z^j), \; z^j(0) = x^j, \quad j = 1, \ldots, m, \; 0 \le t \le 1.$$

Assume that each of these problems has a unique solution $z^j(t, x^j)$ for $0 \le t \le 1$ and any $x^j \in \mathrm{R}^n$. Then, as before, we arrive at a nonlinear system of equations for $x = (x^1, \ldots, x^m)^\top \in \mathrm{R}^N$, $N = n \times m$, of the form $Px + Qz(1, x) - b = 0$, where $z(t, x) = (z^1(t, x^1), \ldots, z^m(t, x^m)) \in \mathrm{R}^N$, $b = (a^\top, 0, \ldots, 0)^\top \in \mathrm{R}^N$, and the matrices $P, Q \in \mathrm{R}^{N \times N}$ are

$$P = \begin{pmatrix} A & 0 & \ldots & 0 \\ 0 & I & \ldots & 0 \\ \vdots & \vdots & \ddots & \vdots \\ 0 & 0 & \ldots & I \end{pmatrix}, \quad Q = \begin{pmatrix} 0 & \ldots & 0 & B \\ -I & \ldots & 0 & 0 \\ \vdots & \ddots & \vdots & \vdots \\ 0 & \ldots & -I & 0 \end{pmatrix}$$

(see, e.g., the monographs by Keller [142], [143] and Ascher, Mattheij, and Russel [12]). Under suitable assumptions, parallel shooting reduces the bound on the error-accumulation in the numerical solution at the expense of an m-fold increase in the dimension of the nonlinear system.

So far, we have considered only infinite-dimensional equations involving differential operators. Clearly, the construction of discrete analogues of other types of equations, such as integral equations, also produces systems of equations in finitely many variables. As a simple example, consider the construction of a conformal mapping from the interior of the complex unit circle $z = \exp(it)$ onto the simply connected domain with the boundary $w = g(r)\exp(ir)$. This problem is essentially solved if we can determine the boundary map $r = r(t)$ which, in turn, is known to satisfy the integral equation of Theodorsen (see, e.g., Gaier [103])

$$r(s) = s + \frac{1}{2\pi} \oint_{-\pi}^{\pi} \log g(r(t)) \, \cot \frac{s-t}{2} \, dt, \tag{2.15}$$

where the Cauchy principal value is taken. A method for approximating the integral by means of trigonometric interpolation goes back to Wittich [280]. The complex trigonometric polynomial

$$p(t) = \sum_{j=-m+1}^{m} \alpha_j e^{ikt}, \quad \alpha_k = \frac{1}{2m} \sum_{j=-m+1}^{m} p_j e^{-ikj\pi/m},$$

satisfies $p(t_k) = p_k$ at the points $t_k = k\pi/m$, $k = -m+1, \ldots, m$. Then, for

$$\bar{p}(s) = s + \frac{1}{2\pi} \oint_{-\pi}^{\pi} p(t) \, \cot \frac{s-t}{2} \, dt,$$

it can be shown (Wittich [280]) that

$$\bar{p}(t_k) = \sum_{j=-m+1}^{m} a_{kj} \, p_j, \quad a_{kj} = \begin{cases} \frac{1}{m} \cot \frac{(k-j)\pi}{2m}, & k-j \text{ even} \\ 0 & j-k \text{ odd.} \end{cases} \tag{2.16}$$

This suggests that with $x = r(t_k)$ and $p_k = \log g(x_k)$, $k = -m+1, \ldots, m$, we approximate (2.15) by the system of equations

$$x = a + A\Phi x,$$

where $a = (t_{-m+1}, \ldots, t_m)^\top \in \mathrm{R}^{2m}$, $A = (a_{kj}) \in L(\mathrm{R}^{2m})$ is the matrix with the elements specified in (2.16), and the nonlinear mapping Φ has the components $\varphi_j(x) = \log g(x_j)$, $j = -m+1, \ldots, m$.

2.2 Minimization

Extremal problems are of foremost importance in almost all applications of mathematics. Many boundary-value problems of mathematical physics may be phrased as variational problems. For instance, holonomic equilibrium problems in Lagrangian mechanics derive from the minimization of a suitable energy function. Similarly, the determination of a geodesic between two points on a manifold is a minimization problem, and so are optimal control problems in engineering, or problems involving the optimal determination of unknown parameters of a technical process. There are close connections between such extremal problems and the solution of nonlinear equations, as is readily seen in the finite-dimensional case.

Let $g : E \subset \mathrm{R}^n \mapsto \mathrm{R}^1$ be some functional. A point $x^* \in E$ is a *local minimizer* of g in E if there exists an open neighborhood $\mathcal{U}$ of x^* in R^n such that

$$g(x) \geq g(x^*), \quad \forall\, x \in \mathcal{U} \cap E, \tag{2.17}$$

and a *global minimizer* on E if the inequality (2.17) holds for all $x \in E$. Moreover, a point x^* in the interior int (E) of E is a *critical point* of g if g has a G-derivative at x^* and $Dg(x^*)^\top = 0$. The relation between local minimizers and critical points is then provided by the following well-known result.

THEOREM 2.1. *For $g : E \subset \mathrm{R}^n \mapsto \mathrm{R}^1$ suppose that $x^* \in \mathrm{int}\,(E)$ is a local minimizer where g is G-differentiable. Then x^* is a critical point of g.*

Of course, a critical point need not be local minimizer. But if g has a second F-derivative at a critical point $x^* \in \mathrm{int}\,(E)$ and $D^2g(x^*)$ is positive-definite then x^* is a proper local minimizer; that is, strict inequality holds in (2.17) for all $x \in \mathcal{U} \cap E$, $x \neq x^*$. Conversely, at a local minimizer x^*, $D^2g(x^*)$ is positive-semidefinite.

The problem of finding critical points of a G-differentiable functional $g : E \subset \mathrm{R}^n \mapsto \mathrm{R}^1$ is precisely that of solving the system of equations

$$Fx = 0, \qquad Fx = Dg(x)^\top, \quad \forall\, x \in E. \tag{2.18}$$

Generally, $F : E \subset \mathrm{R}^n \mapsto \mathrm{R}^n$ is called a *gradient* (or potential) *mapping* on E if there exists a G-differentiable functional $g : E \subset \mathrm{R}^n \mapsto \mathrm{R}^1$ such that $Fx = Dg(x)^\top$ for all $x \in E$. For any gradient mapping the problem (2.18) may be replaced by that of minimizing the functional g, provided, of course, we keep in mind that a local minimizer of g need not be a critical point, nor that a critical point is necessarily a minimizer.

In the continuously differentiable case, gradient mappings can be characterized by the well-known symmetry theorem of Kerner [148].

THEOREM 2.2. *A C^1 map $F : E \subset \mathrm{R}^n \mapsto \mathrm{R}^n$ on an open convex set E is a gradient map on E if and only if $DF(x)$ is symmetric for all $x \in E$.*

This result appears to place a severe limitation on the class of systems which might be solvable by minimizing some nonlinear functional. However, there is always a simple way of converting a system of equations into a minimization problem, and indeed this approach also applies to overdetermined systems.

Consider a mapping $F : E \subset \mathrm{R}^n \mapsto \mathrm{R}^m$, $m \geq n$, and let $f : \mathrm{R}^m \mapsto \mathrm{R}^1$ be a functional which has $x = 0$ as unique global minimizer. For instance, we may choose $f(x) = x^\top Ax$ with a symmetric, positive-definite $A \in L(\mathrm{R}^m)$ or $f(x) = \|x\|$ with some norm on R^m. For given $y \in \mathrm{R}^m$ let

$$(2.19) \qquad g : E \subset \mathrm{R}^n \mapsto \mathrm{R}^1, \quad g(x) = f(Fx - y), \quad \forall\, x \in E,$$

then any solution $x^* \in E$ of $Fx = y$ is certainly a global minimizer of g, and hence we may find x^* by minimizing g. On the other hand, a global minimizer $x^* \in E$ of g need not satisfy $Fx = y$. In fact, this system does not even have to have a solution and, very likely, for $m > n$ it will have none. We call any such global minimizer of g on E an *f-minimal solution* of $Fx = y$.

Various cases of f, such as $f(x) = \|x\|_\infty$ or $f(x) = x^\top x$, are of special interest. For $f(x) = x^\top x$ the functional to be minimized has the form

$$(2.20) \qquad g : E \subset \mathrm{R}^n \mapsto \mathrm{R}^1, \quad g(x) = \sum_{j=1}^{m} (f_j(x) - y_j)^2, \quad \forall\, x \in E,$$

where the f_j are the components of F. This represents a nonlinear least-squares problem. In applications, such least-squares problems may arise, for instance, in the course of estimating certain parameters in a functional relationship on the basis of experimental data (see, e.g., Bates and Watts [21], Björck [30], or Dennis and Schnabel [66]).

As indicated before, an important source of minimization problems is the calculus of variations (see, e.g., Glowinski [110], Struwe [257]). With some continuous function $f : \mathrm{R}^3 \mapsto \mathrm{R}^1$ set

$$(2.21) \qquad J : C^1[0,1] \mapsto \mathrm{R}^1, \quad Ju = \int_0^1 f(s, u(s), u'(s))ds.$$

Then a classical problem of the calculus of variations has the form

$$(2.22) \qquad Ju^* = \inf_{u \in S} Ju, \quad S = \{u \in C^1[0,1] \ : \ u(0) = \alpha, \ u(1) = \beta\}.$$

For instance, in geometric optics a two-dimensional version of Fermat's principle requires that the path $t \mapsto u(t)$ of a light ray between the points $(0, \alpha), (1, \beta) \in \mathrm{R}^2$ is a solution of the variational problem (2.21), (2.22) with

$$f(s, u, p) = \frac{n(s, u)}{c} \sqrt{1 + p^2}.$$

Here n is a positive function on R^2 which defines the index of refraction at the points of the plane and c is the velocity of light in vacuum.

A classical approach to the approximate solution of the variational problem (2.21), (2.22) is the *method of Ritz.* Let $w_i \in C^1[0,1]$, $i = 1, \dots, n$, be linearly independent test functions such that $w_i(0) = w_i(1) = 0$, $i = 1, \dots, n$, and set $\varphi : [0,1] \mapsto R^1$, $\varphi(s) = \alpha + s(\beta - \alpha)$ for $s \in [0,1]$. Then Ritz's method replaces (2.22) by the n-dimensional minimization problem

$$g(x^*) = \inf_{x \in R^n} g(x), \quad g(x) = J\Big(\varphi + \sum_{i=1}^{n} x_i w_i\Big), \quad \forall\, x \in R^n. \tag{2.23}$$

In practice, the integral of J is usually approximated by means of a numerical quadrature formula, say,

$$\int_0^1 k(s)dx \doteq \sum_{j=1}^{m} \gamma_j \; k(s_j). \tag{2.24}$$

Then the approximation of the variational integral (2.21) by the functional

$$\hat{g}(x) = \sum_{j=1}^{m} \gamma_j \; f\Big(s_j, \varphi(s_j) + \sum_{i=1}^{n} x_i w_i(s_j),\; \beta - \alpha + \sum_{i=1}^{n} x_i w_i'(s_j)\Big) \tag{2.25}$$

represents a *discrete Ritz method.* With

$$q : R^{2m} \mapsto R^1, \quad q(y) = \sum_{j=1}^{m} \gamma_j f(s_j, y_{2j-1}, y_{2j}), \quad \forall\, y \in R^{2m}, \tag{2.26}$$

and

$$H = \begin{pmatrix} w_1(s_1) & \dots & w_n(s_1) \\ w_1'(s_1) & \dots & w_n'(s_1) \\ \vdots & \vdots & \vdots \\ w_1(s_m) & \dots & w_n(s_m) \\ w_1'(s_m) & \dots & w_n'(s_m) \end{pmatrix} \in L(R^n, R^{2m}), \quad b = \begin{pmatrix} \varphi(s_1) \\ \beta \quad \alpha \\ \vdots \\ \varphi(s_m) \\ \beta - \alpha \end{pmatrix},$$

the functional $\hat{g}$ of (2.25) can be written in the compact form

$$\hat{g}(x) = q(Hx + b), \quad \forall\, x \in R^n. \tag{2.27}$$

Here q depends only on f and the quadrature formula (2.24), while the affine function, $x \in R^n \mapsto Hx + b$, incorporates the choice of the test functions and the boundary conditions. Properties of $\hat{g}$ can now be derived easily from those of q and hence of f (see, e.g., Stepleman [255]). For instance, if $f = f(s, u, p)$ is convex in (u, p) for fixed s, and if the weights $\gamma_1, \dots \gamma_m$ are positive then (2.26) shows that q is convex, and (2.27) implies the same for $\hat{g}$.

2.3 Discrete Problems

So far, nonlinear systems of equations arose either as finite-dimensional models of infinite-dimensional problems, or as the gradient equations of nonlinear functionals. In a more direct manner, such systems also occur in connection with various naturally finite problems, such as steady-state network-flow problems. This is hardly surprising since one can show, for instance, that the discrete analogue (2.4) of (2.1) may be interpreted as a rectangular *DC* network with the mesh-points as nodes and the mesh segments as conducting, resistive elements.

There is almost as much variety in the types of network problems as in the kinds of, say, elliptic boundary-value problems (see, e.g., Slepian [247]). As before, we restrict ourselves to a simple example.

The basis of the discussion is a finite, directed graph $\Omega = (N, \Lambda)$ with node set $N = \{1, 2, \ldots, p\}$, $p > 0$, and arc set $\Lambda = \{\lambda_1, \ldots, \lambda_q\} \subset N \times N$. Single-node loops $(i, i) \in N \times N$ are excluded from Λ, and the underlying undirected graph Ω shall always be connected, which implies that $q \geq p - 1$. The graph Ω can be described by its node-arc incidence matrix $A \in L(\mathrm{R}^q, \mathrm{R}^p)$ with the elements

$$(2.28) \qquad a_{ij} = \begin{cases} +1 & \text{if } \lambda_j = (i, k) \in \Lambda \text{ for some } k \in N, \\ -1 & \text{if } \lambda_j = (k, i) \in \Lambda \text{ for some } k \in N, \\ 0 & \text{otherwise}, \quad i = 1, \ldots, p, \ j = 1, \ldots, q. \end{cases}$$

Two physical quantities are associated with each arc λ_j, $j = 1, \ldots, q$, namely, a *flow* $y_j \in \mathrm{R}^1$ and a *tension* $u_j \in \mathrm{R}^1$. The specific choices of these variables depend on the particular problem. In a hydraulic network, possible choices for y_j and u_j are the average flow volume in the pipe λ_j and the pressure drop between its ends, respectively. In electrical networks, corresponding choices of variables are the current y_j in λ_j and the voltage drop u_j along that arc.

The vectors $y = (y_1, \ldots, y_q)^\top$ and $u = (u_1, \ldots, u_q)^\top$ are assumed to be connected by a functional relation

$$(2.29) \qquad \Phi(y, u) = 0, \qquad \Phi : \mathrm{R}^q \times \mathrm{R}^q \mapsto \mathrm{R}^q,$$

the *network characteristic*. The specific form of Φ depends on the nature of the network. In the simplest case, each arc $\lambda_j \in \Lambda$ is assumed to be a two-port element defined by a branch characteristic

$$(2.30) \qquad \phi_j(y_j, u_j) = 0, \quad j = 1, \ldots, q,$$

that relates the flow y_j and tension u_j along that specific arc. Thus in this case, we have $\Phi(y, u) = (\phi_1(y_1, u_1), \ldots, \phi_q(y_q, u_q))^\top$.

In the indicated hydraulic network—if turbulent flow prevails—the branch characteristics (2.30) might have the form

$$(2.31) \qquad y_j = \gamma_j \text{ sign } (u_j)|u_j|^\alpha, \quad \gamma_j > 0, \ 0 < \alpha < 1.$$

In an electrical network with branch current y_j and voltage drop u_j along λ_j there are many possible branch characteristics corresponding to the wide variety of two-port devices, such as

current source	$y_j = \text{const}$,
voltage source	$u_j = \text{const}$,
ideal diode	$\max(u_j, -y_j) = 0$,
linear resistor	$y_j = \gamma_j u_j,\ \gamma_j > 0$,
current-driven nonlinear resistor	$u_j = \psi_j(y_j)$,
voltage-driven nonlinear resistor	$y_j = \psi_j(u_j)$.

In addition there are, of course, many k-port devices with $k > 2$, such as all typical transistors. In that case, the network characteristic (2.29) is no longer diagonal.

The law of flow conservation in a network Ω specifies that the (algebraic) sum of the flows over all arcs starting at a node i must equal the sum of the flows over the arcs terminating at i plus the flow b_i supplied at i from the outside. In terms of the node-arc incidence matrix A this means that a flow is any vector $y = (y_1, \ldots, y_q)^\top \in \mathrm{R}^q$ such that $Ay = b$ for given $b = (b_1, \ldots, b_p)^\top \in \mathrm{R}^p$. For $b = 0$, the flow is called a *circulation*, and the arcs carrying nonzero flow form a cycle of Ω. The linear space $\ker A \subset \mathrm{R}^q$ of all circulations is the *cycle space* of Ω. For connected Ω, a basic theorem of graph theory ensures that rank $A = p - 1$ (see, e.g., Slepian [247]), whence, by standard results of linear algebra, $\dim \ker A \equiv r = q - p + 1$. A *cycle matrix* is any matrix $C \in L(\mathrm{R}^r, \mathrm{R}^q)$ for which the columns span $\ker A$, and consequently for which $AC = 0$. A classical algorithm for computing a cycle matrix is based on the construction of a spanning tree of the network graph, (see, e.g., Sedgewick [239], or Seshu and Reed [241]).

A second requirement for our network flow is the conservation of energy, which requires that the (algebraic) sum of the tensions over the arcs of any cycle of Ω should equal the sum of the externally generated tensions on the cycle. In terms of a cycle matrix C this is equivalent to the condition $C^\top u = c$, where $c \in \mathrm{R}^r$ is the vector of the impressed tensions on the basis cycles specified by the columns of C.

Altogether then, a flow problem on Ω requires the determination of $y \in \mathrm{R}^q$ and $u \in \mathrm{R}^q$ such that

$$Ay = b, \quad C^\top u = c, \quad \Phi(y, u) = 0, \tag{2.32}$$

where $b \in \text{rge } A \subset \mathrm{R}^p$ and $c \in \mathrm{R}^r$ are given vectors. This is a system in $2q$ unknowns consisting of q nonlinear and $p + r = q + 1$ linear equations. Because rank $A = p - 1$, the matrix of the linear equations has rank q. There are various ways to reduce the size of (2.32).

Suppose that the characteristic (2.29) can be written as

$$u = \Psi(y). \tag{2.33}$$

Since $b \in \text{rge } A$ there exists $\bar{y} \in \mathrm{R}^q$ such that $A\bar{y} = b$, and, because the columns of C span $\ker A$, it follows that any solution $y \in \mathrm{R}^q$ of $Ay = b$ can be expressed as $y = \bar{y} + Cv$ with a unique $v \in \mathrm{R}^r$. Hence, if $v \in \mathrm{R}^r$ solves

$$C^\top \Psi(\bar{y} + Cv) = c, \tag{2.34}$$

then $y = \bar{y} + Cv$, $u = \Psi(y)$ is a solution of (2.32). The r nonlinear equations (2.34) in the r unknown *mesh currents* $v_1, \ldots, v_r$ are Maxwell's *mesh equations.*

If, instead of (2.33), the network characteristic (2.29) has the form

$$y = \Psi(u), \tag{2.35}$$

then we introduce vectors $x \in \mathrm{R}^p$ of states x_i associated with the nodes $i \in N$, and define any two state vectors x^1, x^2 to be equivalent if $A^\top(x^1 - x^2) = 0$. In other words, we consider the $(p-1)$-dimensional state space $X = \mathrm{R}^p / \ker A^\top$. Since $\ker A^\top = \{x \in \mathrm{R}^p \ : \ x_1 = x_2 = \ldots = x_p\}$, it is simple to work with this space. In fact, it suffices to introduce a boundary condition, e.g., $x_p = 0$, and corresponding partitions

$$A = \begin{pmatrix} A_0 \\ a^\top \end{pmatrix}, \qquad x = \begin{pmatrix} \bar{x} \\ \xi \end{pmatrix}, \qquad \forall\, x \in \mathrm{R}^p,$$

where evidently rank $A_0 = p - 1$. Since rank $C = r$, there exists a $\bar{u} \in \mathrm{R}^q$ such that $C^\top \bar{u} = c$. Then, any solution $u \in \mathrm{R}^q$ of $C^\top u = c$ may be expressed in the form $u = \bar{u} + A_0^\top \bar{x}$ with a unique $\bar{x} \in \mathrm{R}^{p-1}$. Therefore, if $\bar{x} \in \mathrm{R}^{p-1}$ solves

$$A_0 \Psi(\bar{u} + A_0^\top \bar{x}) = \bar{b}, \tag{2.36}$$

then $u = \bar{u} + A_0^\top \bar{x}$, $y = \Psi(u)$ is a solution of (2.32). The $p-1$ equations (2.36) in the $p-1$ states $\bar{x}_1, \ldots, \bar{x}_{p-1}$ represent a special case of *Maxwell's node equations.*

The equations (2.34) and (2.36) are only two examples of the many possible approaches for reformulating the basic problem (2.32), and this flow problem itself admits many important generalizations (see, e.g., Smale [248], and Hachtel, Brayton, and Gustavson [122]).

Our final example is another inherently finite-dimensional problem that leads to a system of nonlinear equations, namely, a simple form of an equilibrium model for a competitive economy. The presentation follows Nikaido [187], and Scarf and Hansen [225].

We consider a pure exchange economy in which n commodities are available for distribution among m economic participants. Each commodity is measured in quantity by a real number $u_i \geq 0$, $i = 1, \ldots, n$. Hence, a "bundle" of commodities is represented by a vector $u \in \mathrm{R}^n$ such that $u \geq 0$ under the componentwise partial ordering on R^n. Any participant has some set $U_j \subset \mathrm{R}^n_+ = \{u \in \mathrm{R}^n \ : \ u \geq 0\}$ of asset vectors of interest and an initial holding defined by a vector $a^j \in U_j$, $j = 1, \ldots, m$.

For any vector $x = (x_1, \ldots, x_n)^\top \geq 0$ of market prices, the jth participant may sell or purchase some assets subject to the constraint that the cost of the purchases equals the income from the sales. Thus $U_j(x) = \{u \in U_j \ : \ x^\top u = x^\top a^j\}$, $j = 1, \ldots, m$, are the sets of assets that will be considered at the price $x \geq 0$. These sets are independent of any scaling of the price vector by some positive factor, and hence we may restrict x to the standard simplex (see also Definition 7.2)

$$\sigma = \left\{ x \in \mathrm{R}^n \ : \ x \geq 0, \ \sum_{i=1}^{n} x_i = 1 \right\}. \tag{2.37}$$

The jth participant is assumed to have established a preference on the consumption set U_j. In other words, for any relative price level $x \in \sigma$ the demand of the jth participant will be any asset vector $u^j \in U_j(x)$ for which the preference rating is maximal. For simplicity we assume here that for any $x \in \sigma$ there is exactly one such vector u^j for the jth consumer, namely, the image value of a known individual demand function

$$(2.38) \qquad \Phi_j : \sigma \subset \mathrm{R}^n \mapsto \mathrm{R}^n, \; \Phi_j(x) \in U_j(x), \quad \forall\, x \in \sigma, \; j = 1, \ldots, m.$$

Then, with $a = a^1 + \ldots + a^m \in \mathrm{R}^n$,

$$(2.39) \qquad H : \sigma \subset \mathrm{R}^n \mapsto \mathrm{R}^n, \quad Hx = a - \sum_{j=1}^{m} \Phi_j(x), \quad \forall\, x \in \mathrm{R}^m,$$

represents the aggregate excess demand function of the economy, and its n components $h_1, \ldots, h_n$ specify the excess demands of the market for the n commodities. The conditions $\Phi_j(x) \in U_j(x)$, $j = 1, \ldots, m$, require the validity of the so-called *Walras law* (in the restricted sense)

$$(2.40) \qquad x^\top Hx = 0, \quad \forall\, x \in \sigma.$$

An equilibrium price vector of the economy is now characterized as any $x^* \in \sigma$ for which $Hx^* \leq 0$; that is, for which the excess demand $h_j(x^*)$ of the market for any commodity i is nonpositive. From (2.40) it follows then that $h_i(x^*) = 0$ whenever $x_i^* > 0$.

The consumption sets are usually unbounded and hence it may happen that the excess aggregate demand for a particular commodity tends to infinity when the price goes to zero. In general, this is economically unrealistic, and can be avoided by allowing each consumer to maximize the preferences only over $U_j(x) \cap \{u : b^j \leq u - a^j \leq c^j\}$ with suitably chosen finite bounds $b^j, c^j \in \mathrm{R}^n$, $j = 1, \ldots, n$. With this it is then reasonable to assume that the aggregate excess demand function $H : \sigma \mapsto \mathrm{R}^n$ is a continuous mapping on all of σ with a bounded range. Now the function

$$G : \sigma \subset \mathrm{R}^n \mapsto \mathrm{R}^n, \; g_i(x) = \frac{x_i + \max(0, h_i(x))}{1 + \sum_{i=1}^n \max(0, h_i(x))}, \; \forall x \in \sigma, i = 1, \ldots, n,$$

is well defined and continuous on σ and evidently we have $Gx \in \sigma$ for any $x \in \sigma$. Without further discussion, we note that, by the Brouwer fixed-point theorem (see Theorem 10.2), this implies the existence of a fixed point $x^* = Gx^*$, $x^* \in \sigma$. If, say,

$$c = 1 + \sum_{i=1}^{n} \max(0, h_i(x^*)) > 1,$$

then it follows from $x_i^* + \max(0, h_i(x^*)) = cx_i^*$, $i = 1, \ldots, n$, that $h_i(x^*) > 0$ whenever $x_i^* > 0$. By (2.40) this is impossible and therefore we have $c = 1$ and $\max(0, h_i(x^*)) = 0$, $i = 1, \ldots, n$. Thus we have $Hx^* \leq 0$ which shows that

x^* is an equilibrium price vector and that, under our assumptions, the pure exchange economy discussed here always possesses an equilibrium price level. This is a special case of a theorem on Walrasian competitive economies (see Nikaido [187] for a discussion and further references). At the same time, we see that in our setting the problem of finding the equilibrium price vector is equivalent to that of solving an n-dimensional nonlinear fixed-point equation on the standard simplex. In a more general case, a fixed-point problem for a set-valued mapping has to be considered as discussed, e.g., by Aubin [13].

CHAPTER 3

Iterative Processes and Rates of Convergence

Iterative processes for the solution of finite-dimensional nonlinear equations vary almost as widely in form and properties as do the equations themselves. In this chapter we introduce an algorithmic characterization of a class of these processes and then discuss some measures of their efficiency and rate of convergence.

3.1 Characterization of Iterative Processes

Let $F : E \subset \mathrm{R}^n \mapsto \mathrm{R}^n$ be a given mapping and suppose that a zero of F; that is, a solution of the equation $Fx = 0$, is to be computed. As noted in section 1.1 it is, in general, not certain that such a solution exists, nor is the number of such solutions obvious. In this chapter we simply assume that a particular zero x^* exists and is to be computed.

In contrast to systems of linear equations, an explicit solution of nonlinear systems of equations is rarely feasible. Consequently, attention will be restricted to iterative methods. Any such iterative process, to be denoted here by $\mathcal{J}$, has two principal parts, namely an algorithm $\mathcal{G}$ constituting a single iteration step, and an acceptance algorithm $\mathcal{A}$ to control the iteration. We define the input of both $\mathcal{G}$ and $\mathcal{A}$ to be a triple $\{k, x, M\}$ consisting of an *iteration index* k, the current *iterate* $x \in \mathrm{R}^n$, and a *memory set* M containing certain known process data other than x and k. The output $step = \mathcal{G}(k, x, M)$ of $\mathcal{G}$ is either an error return $step =$ "fail" or a pair $step = \{\bar{x}, \bar{M}\}$ consisting of the next iterate $\bar{x}$ and the next memory set $\bar{M}$. The output of $\mathcal{A}$ is a variable that has one of the three values "accept", "continue", "fail" signifying a successful completion of the iteration, the need for another step, or a fatal error condition, respectively. With this the iterative process $\mathcal{J}$ is an algorithm of the generic form (3.1).

In general, the algorithms $\mathcal{G}$ and $\mathcal{A}$ depend on the mapping F; that is, they may have to incorporate calls to all necessary procedures for evaluating F or, if needed, any of its derivatives. The memory set contains data computed in prior steps or varying with the input to $\mathcal{J}$, such as several or all previous iterates, or certain retained function values. The specific contents of the memory set depend on the particular implementation, but usually redundant data are retained whenever the recomputation is costly. With a somewhat different meaning, the term "memory" of an iterative process was introduced by Traub [265].

For the theoretical analysis the acceptance algorithm $\mathcal{A}$ is rendered inactive;

that is, $\mathcal{A}$ is assumed always to return the value "continue". Hence, unless the step routine fails for some k, the process produces an infinite sequence of iterates (which, of course, need not converge). For the remainder of the chapter we assume, unless otherwise specified, that $\mathcal{A}$ is inactive.

(3.1)

```
J: input: {x, M}
   k := 0; decision := continue;
   while decision = continue
      output: {k, x, M};
      decision := A{k, x, M};
      if decision = accept then
            return: "iteration completed";
      if k > k_max ∧ decision = fail then
            return: "iteration failed";
      step := G{k, x, M};
      if step = fail then
            return: "step failed";
      {x, M} := step; k := k + 1;
   endwhile
```

We call a process $\mathcal{J}$ *stationary* if the output of $\mathcal{G}$ does not depend on the current value of the iteration index k. This means that, if $\mathcal{G}$ returns $\{x^j, M_j\}$ at step $j > 0$, and we restart $\mathcal{J}$ with $\{\bar{x}^0, \bar{M}_0\}$ where $\bar{x}^0 = x^j$, and $\bar{M}_0 = M_j$, then the resulting sequence $\{k, \bar{x}^k, \bar{M}_k\}$ satisfies $\bar{x}^k = x^{j+k}$, $\bar{M}_k = M_{j+k}$. Otherwise, $\mathcal{J}$ is *nonstationary*. The process $\mathcal{J}$ is an m-step method, $m \geq 1$, if for any $k \geq m$ the step algorithm $\mathcal{G}$ depends exactly on m prior iterates for the computation of its output. This necessitates, of course, some modification of $\mathcal{G}$ for all steps with $k < m$. Clearly, the required prior iterates have to be contained in the memory set M_k. If, in particular, for $k \geq m$, $\mathcal{G}$ depends exactly on $x^k, x^{k-1}, \ldots, x^{k-m+1}$, then we speak of a sequential m-step method. In the simple case $m = 1$ we have a *one-step* process. If, in addition, $\mathcal{J}$ is stationary, then $\mathcal{G}$ defines a mapping $G : D_G \subset \mathrm{R}^n \mapsto \mathrm{R}^n$ and a step of the method can be written in the familiar form $x^{k+1} = Gx^k$, $k = 0, 1, \ldots$.

The input $\{x^0, M_0\}$ to the process $\mathcal{J}$ is admissible if the step algorithm $\mathcal{G}$ never fails and hence $\mathcal{J}$ produces an infinite sequence of triples $\{k, x^k, M_k\}$; that is, in particular, a *sequence of iterates* $\{x^k\} \subset \mathrm{R}^n$. The set of all admissible inputs is the product of a set $\mathrm{PD}(\mathcal{J}) \subset \mathrm{R}^n$ of admissible starting points x^0, and a set $\mathrm{MD}(\mathcal{J})$ of admissible memory sets M_0 in an unspecified space. Often, for a specific process and equation, an admissible input is fully characterized by $x^0 \in \mathrm{PD}(\mathcal{J})$; but there are many examples, including most methods of secant type (see chapter 5) where x^0 alone is not sufficient to specify an admissible input.

Assume that the set of admissible inputs is not empty. A point $x^* \in \mathrm{R}^n$ is a limit point of $\mathcal{J}$ if there exists a sequence of iterates $\{x^k\}$ with $\lim_{k\to\infty} x^k = x^*$. The set of all sequences of iterates $\{x^k\}$ which are generated by $\mathcal{J}$ and converge to x^* is denoted by $C(\mathcal{J}, x^*)$. It is, of course, rarely possible to determine $C(\mathcal{J}, x^*)$ or, for that matter, the corresponding set of all admissible inputs of

$\mathcal{J}$. Hence, the convergence analysis of $\mathcal{J}$ will aim at determining sufficient conditions for admissible inputs such that the corresponding sequences of iterates are guaranteed to belong to $C(\mathcal{J}, x^*)$ for some solution x^* of $Fx = 0$.

Of course, for any practical application of $\mathcal{J}$ a suitable acceptance test $\mathcal{A}$ has to be used in $\mathcal{J}$. Such a test should take into account the influence of roundoff when $\mathcal{J}$ is executed on a computer. Let $\{x^k\} \subset C(\mathcal{J}, x^*)$ be a sequence of iterates generated by $\mathcal{J}$ (without active $\mathcal{A}$ and in real arithmetic) and $\{x_d^k\}$ the corresponding output of a machine implementation of $\mathcal{J}$ using some d-digit arithmetic. Suppose that when $\mathcal{A}$ is active the process terminates the sequence $\{x_d^k\}$ with the iterate $x_d^{k^*}$. Then we might demand of a satisfactory acceptance test that

$$\lim_{d\to\infty} x_d^{k^*} = x^*. \tag{3.2}$$

As a typical example, consider a frequently used test which terminates the process at the first index $k^* = k^*(d)$ with the property that

$$\|x_d^{k^*+1} - x_d^{k^*}\| \le \epsilon_d \|x_d^{k^*}\|,$$

where the given tolerances $\epsilon_d > 0$ satisfy $\lim_{d\to\infty} \epsilon_d = 0$. For this test (3.2) holds if it is known that

$$\|x^{k+1} - x^k\| \le \alpha\|x^k - x^{k-1}\|, \quad \|x_d^k - x^k\| \le \rho_d, \quad k = 1, 2, \ldots,$$

with a fixed $\alpha < 1$ and roundoff bounds that satisfy $\lim_{d\to\infty} \rho_d = 0$. In fact, since by assumption $\lim_{k\to\infty} x^k = x^*$, we obtain

$$\begin{aligned}\|x^{k^*} - x^*\| &\le \sum_{j=k^*}^{\infty} \|x^{j+1} - x^j\| \\ &\le \frac{1}{1-\alpha}\left[\|x_d^{k^*+1} - x_d^{k^*}\| + 2\rho_d\right] \le \frac{1}{1-\alpha}\left[\epsilon_d\|x_d^{k^*}\| + 2\rho_d\right]\end{aligned}$$

or

$$\|x_d^{k^*} - x^*\| \le \rho_d + \frac{1}{1-\alpha}\left[\epsilon_d\|x^{k^*}\| + \rho_d(2 + \epsilon_d)\right],$$

which proves the assertion.

Clearly, the design of any practical acceptance test depends strongly on the problem class, the theoretical properties of $\mathcal{J}$, the information available when $\mathcal{A}$ is performed, the implementation of $\mathcal{J}$, and the characteristics of the particular machine. It is therefore not surprising that most known general results about such tests, as that of the example, are proved under rather stringent assumptions. Many different types of tests have been proposed and used; at the same time, several examples of Nickel and Ritter [186] show that already in simple cases some of the "reasonable" tests may fail the above criteria. This points to the necessity for tailoring an acceptance test as much as possible to the specific situation at hand and to consider proving its satisfactory behavior only in that setting. Few results along this line appear to be available in the literature.

3.2 Rates of Convergence

As noted in the introduction, a basic aspect of any iterative process is its computational complexity, which includes questions not only about the cost of executing the process, but also about the characterization of optimal members in a class of comparable methods. We restrict ourselves here to some basic results concerning the first of these questions.

If the cost of one basic iteration step of a process $\mathcal{J}$ is essentially constant, the overall effort of executing $\mathcal{J}$ is proportional to the number of steps required to reach an acceptable iterate. For any sequence of iterates $\{x^k\} \in C(\mathcal{J}, x^*)$ this poses the question of when the absolute errors $\epsilon_k = \|x^k - x^*\|$, $k = 0, 1, \ldots$ will remain below a given tolerance. In other words, we are interested in the rate of convergence of the real, nonnegative sequence $\{\epsilon_k\}$ with limit zero. The search for suitable measures of the convergence rate of such sequences is probably as old as convergence theory itself. We summarize here an approach given in [OR] which was modeled on the standard quotient and root tests for infinite series; further references may also be found there.

DEFINITION 3.1. *For a sequence $\{x^k\} \subset \mathrm{R}^n$ with limit $x^* \in \mathrm{R}^n$ and any $p \in [1, \infty)$, the numbers*

$$(3.3) \qquad R_p\{x^k\} = \begin{cases} \limsup\limits_{k\to\infty} \|x^k - x^*\|^{1/k}, & \text{if } p = 1, \\ \limsup\limits_{k\to\infty} \|x^k - x^*\|^{1/p^k}, & \text{if } p > 1, \end{cases}$$

are the root-convergence factors (R-factors) of $\{x^k\}$. Moreover,

$$(3.4) \qquad R_p(\mathcal{J}, x^*) = \sup\{\, R_p\{x^k\} \;:\; \{x^k\} \in C(\mathcal{J}, x^*) \,\}, \quad \forall\, p \in [1, \infty),$$

are the R-factors of an iterative process $\mathcal{J}$ with limit point x^*.

DEFINITION 3.2. *For a sequence $\{x^k\} \subset \mathrm{R}^n$ with limit $x^* \in \mathrm{R}^n$ and any $p \in [1, \infty)$, the quantities*

$$(3.5) \qquad Q_p\{x^k\} = \begin{cases} \limsup\limits_{k\to\infty} \dfrac{\|x^{k+1} - x^k\|}{\|x^k - x^*\|^p}, & \text{if } x^k \neq x^* \ \forall\, k \geq k_0, \\ 0, & \text{if } x^k = x^* \ \forall\, k \geq k_0, \\ +\infty, & \text{otherwise}, \end{cases}$$

are the quotient-convergence factors (Q-factors) of $\{x^k\}$ with respect to the particular norm. Moreover,

$$(3.6) \qquad Q_p(\mathcal{J}, x^*) = \sup\{\, Q_p\{x^k\} \;:\; \{x^k\} \in C(\mathcal{J}, x^*) \,\}, \quad \forall\, p \in [1, \infty),$$

are the Q-factors of an iterative process $\mathcal{J}$ with limit x^*.

Clearly, whenever they are defined for a sequence or process, the Q-factors belong to $[0, \infty)$ and the R-factors to $[0, 1]$. Simple examples show that the Q-factors may depend on the norm while the equivalence of all norms on R^n implies the norm-independence of the R-factors. In fact, let $\|\cdot\|_a$, and $\|\cdot\|_b$ denote two norms on R^n and consider any sequence $\{\gamma_k\}$ of real positive numbers converging to zero. Then, by using (1.2) we obtain

$$\begin{aligned}\limsup_{k\to\infty} \|x^k - x^*\|_a^{\gamma_k} &\le \lim_{k\to\infty} c_2^{\gamma_k} \limsup_{k\to\infty} \|x^k - x^*\|_b^{\gamma_k} \\ &= \limsup_{k\to\infty} \|x^k - x^*\|_b^{\gamma_k} \le \lim_{k\to\infty} \frac{1}{c_1}^{\gamma_k} \limsup_{k\to\infty} \|x^k - x^*\|_a^{\gamma_k} \\ &= \limsup_{k\to\infty} \|x^k - x^*\|_a^{\gamma_k},\end{aligned}$$

which implies that $R_p\{x^k\}$ and therefore also $R_p(\mathcal{J}, x^*)$ are norm-independent.

A central fact is that, as functions of p, these factors are nondecreasing step functions with at most one step. This is the content of the following two theorems.

THEOREM 3.1. *For any iterative process $\mathcal{J}$ with limit x^* exactly one of the following conditions holds:*

(a) $R_p(\mathcal{J}, x^*) = 0$ for $p \in [1, \infty)$;

(b) $R_p(\mathcal{J}, x^*) = 1$ for $p \in [1, \infty)$;

(c) there exists a $p_0 \in [1, \infty)$ such that $R_p(\mathcal{J}, x^*) = 0$ for $p \in [1, p_0)$ and $R_p(\mathcal{J}, x^*) = 1$ for $p \in [1, p_0)$.

Proof. Let $\{x^k\}$ be any sequence converging to x^* and set $\gamma_{1k} = 1/k$ for $p = 1$ and $\gamma_{pk} = 1/p^k$ for $p > 1$ and all $k \ge 0$. Then the quotients γ_{qk}/γ_{pk} tend to infinity whenever $1 \le q < p$. Suppose that $R_p\{x^k\} < 1$ for some $p \in (1, \infty)$ and select $\epsilon > 0$ such that $R_p\{x^k\} + \epsilon = \alpha < 1$. Set $\epsilon_k = \|x^k - x^*\|$ and choose $k \ge k_0$ such that $\epsilon_k^{\gamma_{pk}} \le \alpha$ for all $k \ge k_0$. Then, for any $q \in [1, p)$ we obtain

$$R_q\{x^k\} = \limsup_{k\to\infty} (\epsilon_k^{\gamma_{pk}})^{\gamma_{qk}/\gamma_{pk}} \le \lim_{k\to\infty} \alpha^{\gamma_{qk}/\gamma_{pk}} = 0;$$

that is, $R_q\{x^k\} = 0$ whenever $q < p$ and $R_p\{x^k\} < 1$. This also shows that $R_q\{x^k\} = 1$ whenever $q > p$ and $R_p\{x^k\} > 0$. Hence, one and only one of the three conditions (a), (b), (c) can hold. Suppose that neither (a) nor (b) holds. Then $p_0 = \inf\{p \in [1, \infty) : R_p(\mathcal{J}, x^*) = 1\}$ is well defined. If there exists $p > p_0$ such that $R_p(\mathcal{J}, x^*) < 1$, then $R_p\{x^k\} < 1$ for all $\{x^k\} \in C(\mathcal{J}, x^*)$, while, by definition of p_0, there exists a $p' \in [p_o, p)$ such that $R_{p'}(\mathcal{J}, x^*) = 1$. In particular, therefore, $R_{p'}\{x^k\} > 0$ for some sequence in $C(\mathcal{J}, x^*)$, which, by the first part of the proof, implies that $R_p\{x^k\} = 1$ for this sequence; this is a contradiction. Thus $R_p(\mathcal{J}, x^*) = 1$ for $p = p_0$, and, similarly, it follows that $R_p(\mathcal{J}, x^*) = 0$ for $p < p_0$. □

THEOREM 3.2. *For any iterative process $\mathcal{J}$ with limit x^* exactly one of the following conditions holds:*

(a) $Q_p(\mathcal{J}, x^*) = 0$ for $p \in [1, \infty)$;

(b) $Q_p(\mathcal{J}, x^*) = \infty$ for $p \in [1, \infty)$;

(c) There exists $p_0 \in [1, \infty)$ such that $Q_p(\mathcal{J}, x^*) = 0$ for $p \in [1, p_0)$, and $Q_p(\mathcal{J}, x^*) = \infty$ for $p \in (p_0, \infty)$.

Moreover, the three relations $Q_p(\mathcal{J}, x^*) = 0$, $0 < Q_p(\mathcal{J}, x^*) < \infty$, and $Q_p(\mathcal{J}, x^*) = \infty$ are independent of the norm.

The proof is analogous to that of Theorem 3.1 and will be omitted. For such a proof see section 9.1 of [OR].

The values of p_0 at which the steps occur in the two theorems are the corresponding orders of convergence of the following process.

DEFINITION 3.3. *Let $\mathcal{J}$ be an iterative process with limit x^*. Then the R-order of $\mathcal{J}$ at x^* is*

$$O_R(\mathcal{J}, x^*) = \inf\{p \ : \ 1 \le p < \infty,\ R_p(\mathcal{J}, x^*) = 1\} \tag{3.7}$$

unless $R_p(\mathcal{J}, x^*) = 0$ for all $p \ge 1$ in which case $O_R(\mathcal{J}, x^*) = \infty$. Analogously, the Q-order of $\mathcal{J}$ at x^* is

$$O_Q(\mathcal{J}, x^*) = \inf\{p \ : \ 1 \le p < \infty,\ Q_p(\mathcal{J}, x^*) = \infty\}, \tag{3.8}$$

unless $Q_p(\mathcal{J}, x^*) = 0$ for all $p \ge 1$, in which case $O_Q(\mathcal{J}, x^*) = \infty$.

These orders are both norm-independent; in fact, for the Q-order this follows from Theorem 3.2, while for the R-order it is a consequence of the norm-independence of the R-factors. Note that for any process $\mathcal{J}$ the convergence factors, and hence also the orders, depend on the limit point x^*. For a different limit point these quantities may well differ.

The Q- and R-factors can be used to compare the rate of convergence of different iterative processes. A comparison of two iterative processes, $\mathcal{J}_1$ and $\mathcal{J}_2$, in terms of the R-measure proceeds as follows. First compare the R-orders $o_i = O_R(\mathcal{J}_i, x^*)$, $i = 1, 2$; the process with the larger R-order is R-faster than the other one. For $o_1 = o_2 = p$ compare the R-factors $\gamma_i = R_p(\mathcal{J}_i, x^*)$, $i = 1, 2$; if one of these numbers is smaller, the corresponding process is R-faster. For the Q-measure we proceed analogously, except that now there is a possible norm-dependence. If the Q-orders $o_1 = O_Q(\mathcal{J}_1, x^*)$ and $o_2 = O_Q(\mathcal{J}_2, x^*)$ are different, then the process with the larger Q-order is Q-faster than the other one under any norm. For $o_1 = o_2 = p$ we compute the Q-factors $\gamma_1 = Q_p(\mathcal{J}_1, x^*)$ and $\gamma_2 = Q_p(\mathcal{J}_2, x^*)$, of course, in the same norm. If, say, $0 = \gamma_1 < \gamma_2$ or $\gamma_1 < \gamma_2 = \infty$, then $\mathcal{J}_1$ is Q-faster than $\mathcal{J}_2$ no matter which norm is used. However, for $0 < \gamma_1 < \gamma_2 < \infty$ we can only say that $\mathcal{J}_1$ is Q-faster than $\mathcal{J}_2$ under the particular norm, and the situation may reverse itself under another one.

Iterative processes of order at most two are most frequent. The convergence is said to be R-quadratic or Q-quadratic if $O_R(\mathcal{J}, x^*) = 2$ or $O_Q(\mathcal{J}, x^*) = 2$, respectively. In the cases $R_1(\mathcal{J}, x^*) = 0$, $0 < R_1(\mathcal{J}, x^*) < 1$, or $R_1(\mathcal{J}, x^*) = 1$ we speak of R-superlinear, R-linear, or R-sublinear convergence, respectively. Correspondingly the relations $Q_1(\mathcal{J}, x^*) = 0$, $0 < Q_1(\mathcal{J}, x^*) < 1$, and $Q_1(\mathcal{J}, x^*) \ge 1$ define Q-superlinear, Q-linear, and Q-sublinear convergence, respectively. Except for the last two concepts, all others are norm independent.

The following result provides a useful characterization of Q-superlinear convergence.

THEOREM 3.3. *Any sequence $\{x^k\} \subset \mathrm{R}^n$ with $\lim_{k\to\infty} x^k = x^*$ satisfies*

$$\left| 1 - \frac{\|x^{k+1} - x^k\|}{\|x^k - x^*\|} \right| \le \frac{\|x^{k+1} - x^*\|}{\|x^k - x^*\|}, \qquad \textit{whenever } x^k \ne x^*.$$

Hence, if the convergence is Q-superlinear and $x^k \neq x^*$ for all large k, then

$$\lim_{k\to\infty} \frac{\|x^{k+1} - x^k\|}{\|x^k - x^*\|} = 1.$$

The proof follows directly from the inequalities

$$\Big| \, \|x^* - x^k\| - \|x^{k+1} - x^*\| \, \Big| \leq \|x^{k+1} - x^k\| \leq \|x^* - x^k\| + \|x^{k+1} - x^*\|.$$

As a final result we mention some basic relations between the Q- and R-orders and the Q_1- and R_1-factors but refer for a proof to [OR].

THEOREM 3.4. *For any iterative process $\mathcal{J}$ with limit x^*, and under any norm, we have $O_Q(\mathcal{J}, x^*) \leq O_R(\mathcal{J}, x^*)$ and $R_1(\mathcal{J}, x^*) \leq Q_1(\mathcal{J}, x^*)$.*

3.3 Evaluation of Convergence Rates

The practical value of any complexity measure depends on the ease with which it may be estimated for specific processes. It is therefore a point in favor of the Q- and R-rates of convergence that for them some general characterization theorems are available for certain classes of methods. We begin with the following concept characterizing local convergence.

DEFINITION 3.4. *For given $G : E \subset \mathrm{R}^n \mapsto \mathrm{R}^n$, a point of attraction of the stationary one-step process*

$$\mathcal{J}: \quad x^{k+1} = Gx^k, \quad k = 0, 1, \ldots, \tag{3.9}$$

is any point $x^* \in E$ which possesses an open neighborhood $\mathcal{U} \subset E$ with the property that for any $x^0 \in \mathcal{U}$ the iterates $\{x^k\}$ of (3.9) are well defined, remain in E, and converge to x^*.

If G is continuous at x^* then, clearly, $x^* = Gx^*$ is a fixed point of G. When G is affine, say, $Gx = Bx + b$ with given $B \in L(\mathrm{R}^n)$ and $b \in \mathrm{R}^n$, then the iterates $\{x^k\}$ of (3.9) converge to the unique fixed point x^* of G in R^n starting from any $x^0 \in \mathrm{R}^n$ if and only if B has spectral radius $\rho(B) < 1$ (see, e.g., Golub and Van Loan [114], Varga [269], or Young [284]). Moreover, in that case, we have $R_1(\mathcal{J}, x^*) = \rho(B)$. The corresponding theorem in the nonaffine case is unfortunately not as general.

THEOREM 3.5. *Suppose that $G : E \subset \mathrm{R}^n \mapsto \mathrm{R}^n$ has a fixed point $x^* \in \mathrm{int}\,(E)$ where G is F-differentiable. If $\sigma = \rho(DG(x^*)) < 1$, then x^* is a point of attraction of the process $\mathcal{J}$ defined by (3.9) and $R_1(\mathcal{J}, x^*) = \sigma$. Moreover, if $\sigma > 0$, then $O_R(\mathcal{J}, x^*) = O_Q(\mathcal{J}, x^*) = 1$.*

Proof. For $\epsilon > 0$ with $\sigma + 2\epsilon < 1$ there exists a norm such that $\|DG(x^*)\| \leq \sigma + \epsilon$ and a ball $\mathcal{B} = B(x^*, \delta) \subset E$, $\delta > 0$, such that

$$\|Gx - Gx^* - DG(x^*)(x - x^*)\| \leq \epsilon \|x - x^*\| \quad \forall\, x \in \mathcal{B}. \tag{3.10}$$

Then for all $x \in \mathcal{B}$ it follows that

$$\begin{aligned} \|Gx - x^*\| &\leq \|Gx - Gx^* - DG(x^*)(x - x^*)\| \\ &\qquad + \|DG(x^*)\| \, \|x - x^*\| \\ &\leq (\sigma + 2\epsilon)\|x - x^*\|. \end{aligned} \tag{3.11}$$

Hence, for $x^0 \in \mathcal{B}$ induction shows that, for $k = 0, 1, \ldots,$

$$\|x^{k+1} - x^*\| \leq (\sigma + 2\epsilon)\|x^k - x^*\| \leq \ldots \leq (\sigma + 2\epsilon)^{k+1}\|x^0 - x^*\|,$$

which proves the local convergence. At the same time, this inequality also shows that $R_1(\mathcal{J}, x^*) \leq \sigma + 2\epsilon$ and, therefore, because of the norm-independence of the R-factors, that $R_1(\mathcal{J}, x^*) \leq \rho(DG(x^*))$.

For $\sigma = \rho(DG(x^*)) = 0$ the proof is complete; thus assume that $\sigma > 0$. Let $\lambda_1, \ldots, \lambda_n \in \mathrm{C}^n$ be the eigenvalues of $DG(x^*)$ ordered so that $|\lambda_i| = \sigma$, $i = 1, \ldots, m$, $|\lambda_i| < \sigma$, $i = m+1, \ldots, n$. Set $\alpha = \max\{|\lambda_i| \,:\, i = m+1, \ldots, n\}$ or $\alpha = 0$ if $m = n$. Clearly, $\alpha < \sigma$, and we may select $\epsilon > 0$ such that $\alpha + 3\epsilon < \sigma - 3\epsilon$, $\sigma + 2\epsilon < 1$. Now choose a basis $u^1, \ldots, u^n$ of C^n such that in this basis, $DG(x^*)$ has the Jordan form

$$DG(x^*) = \begin{pmatrix} \lambda_1 & \epsilon_1 & \cdots & 0 \\ \vdots & \ddots & \ddots & \vdots \\ 0 & 0 & \ddots & \epsilon_{n-1} \\ 0 & 0 & \cdots & \lambda_n \end{pmatrix} \tag{3.12}$$

where the ϵ_i are either 0 or ϵ, and, in particular, $\epsilon_m = 0$. Then any $y \in \mathrm{R}^n$ may be expressed as $y = \eta_1 u^1 + \cdots + \eta_n u^n$ where, in general, the coefficients are complex and $y \in \mathrm{R}^n \mapsto \|y\|_u = |\eta_1| + \cdots + |\eta_n|$ defines a norm on R^n, and in the induced matrix norm we have $\|DG(x^*)\|_u \leq \sigma + \epsilon < 1$. Hence, from the first part of the proof it follows that, for any $x^0 \in \mathcal{B}$, we have $\{x^k\} \subset \mathcal{B}$ and $\lim_{k\to\infty} x^k = x^*$. Set

$$y^k = x^k - x^*, \ \forall\, k \geq 0, \quad \gamma_k = \sum_{i=1}^{m} |\eta_i^k|, \ \beta_k = \sum_{i=m+1}^{n} |\eta_i^k|,$$

where $\eta_1^k, \ldots, \eta_n^k$ are the coefficients of y^k in the basis $u^1, \ldots, u^n$. Choose $x^0 \in \mathcal{B}$ such that $\beta_0 < \gamma_0$. We show by induction that $\beta_p < \gamma_p$ for all $p \geq 0$. Assume that this holds for $p = 0, \ldots, k$. Then $y^{k+1} = DG(x^*)y^k + R(x^k)$ where $R(x) = Gx - Gx^* - DG(x^*)(x - x^*)$, and it follows from (3.12) that

$$\begin{aligned} \beta_{k+1} &= \sum_{i=m+1}^{n} |\eta_i^{k+1}| = \sum_{i=m+1}^{n} |\lambda_i \eta_i^k + \epsilon_i \eta_i^{k+1} + r_i(x^k)| \\ &\leq (\alpha + \epsilon)\beta_k + \|R(x^k)\|_u. \end{aligned}$$

Therefore, since $\|y^k\|_u = \gamma_k + \beta_k$, it follows from (3.10) and the induction hypothesis that

$$\beta_{k+1} \leq (\alpha + \epsilon)\beta_k + \epsilon(\gamma_k + \beta_k) \leq (a + 3\epsilon)\gamma_k.$$

Similarly, recalling $\epsilon_m = 0$, we have

$$\begin{aligned} \gamma_{k+1} &= \sum_{i=1}^{m} |\lambda_i \eta_i^k + \epsilon_i \eta_i^{k+1} + r_i(x^k)| \\ &\geq \sigma\gamma_k - \epsilon\gamma_k - \epsilon(\gamma_k + \beta_k) \geq (\sigma - 3\epsilon)\gamma_k. \end{aligned} \tag{3.13}$$

Hence, the choice of ϵ implies that $\beta_{k+1} \leq [(\alpha + 3\epsilon)/(\sigma - 3\epsilon)]\gamma_{k+1} < \gamma_{k+1}$, which completes the induction. Therefore, it also follows that (3.13) holds for all $k \geq 0$, and hence

$$\|x^k - x^*\|_u = \gamma_k + \beta_k \geq \gamma_k \geq (\sigma - 3\epsilon)\gamma_{k-1} \geq \cdots \geq (\sigma - 3\epsilon)^k \gamma_0.$$

This shows that $R_1\{x^k\} \geq \sigma - 3\epsilon$, and, since ϵ was arbitrary, that $R_1\{x^k\} \geq \sigma$, so that $R_1(\mathcal{J}, x^*) \geq \sigma$. Together with the first part of the proof this proves that $R_1(\mathcal{J}, x^*) = \sigma$. Finally, if $\sigma \neq 0$, then, by Definition 3.1 we have $O_R(\mathcal{J}, x^*) = 1$ and, by Theorem 3.4, $Q_1(\mathcal{J}, x^*) > 0$. Moreover, the first part of the proof exhibited a norm in which $Q_1(\mathcal{J}, x^*) < 1$. Together this implies that $O_Q(\mathcal{J}, x^*) = 1$. □

The first part of the proof of Theorem 3.5 is analogous to that in the affine case. But, in contrast to that case, note that the existence of x^* had to be assumed, only local instead of global convergence is guaranteed, and merely the sufficiency but not the necessity of the condition $\sigma < 1$ is asserted. Appropriate counterexamples show that under the assumptions of Theorem 3.5 no improvement may be expected. For instance, in the one-dimensional case $Gx = x - x^3$ we have local convergence at $x^* = 0$ even though $\rho(DG(x^*)) = 1$, while for $Gx = x + x^3$ the point $x^* = 0$ is no longer a point of attraction. For various other related examples, see Voigt [270] and [OR].

The sufficiency of $\rho(DG(x^*)) < 1$ for a point of attraction was first proved by Ostrowski [196] under somewhat more stringent conditions on G, and later by Ostrowski [197], and Ortega and Rockoff [194], under those of Theorem 3.5. However, in essence, this result had already been obtained by Perron [202]. In fact, for the perturbed linear difference equation

$$z^{k+1} = Az^k + \Phi(z^k), \quad k = 0, 1, \ldots, \quad A \in L(\mathbb{R}^n), \tag{3.14}$$

Perron showed that if $\rho(A) < 1$ and $\Phi : B(0, \delta) \subset \mathbb{R}^n \mapsto \mathbb{R}^n$ is small in the sense that $\lim_{\|x\| \mapsto 0} \Phi(x)/\|x\| = 0$, then for sufficiently small $\|z^0\|$ the sequence $\{z^k\}$ converges to zero. In our case, we have $DG(x^*) = A$, $z^k = x^k - x^*$ and $\Phi(x) = Gx - Gx^* - DG(x^*)(x - x^*)$. The second part of the proof asserting the equality $R_1(\mathcal{J}, x^*) = \rho(DG(x^*)) = \sigma$ was given in [OR] and is an adaptation of a proof of Coffmann [51] on the asymptotic behavior of solutions of perturbed linear difference equations.

As these remarks indicate, the local convergence of iterative processes represents a result about the asymptotic behavior of the solutions of difference equations at a stationary point. Therefore, they also bear some relation to the large body of results on the asymptotic behavior of the solutions of differential equations near stationary points.

Theorem 3.5 gives the rate of convergence only in the R-measure. Suppose, however, that $\|DG(x^*)\| = \sigma$ under some norm. Then, the first inequality in (3.11) together with the definition of F-differentiability shows that

$$\lim_{x \to \infty} \frac{\|Gx - Gx^*\|}{\|x - x^*\|} \leq \sigma. \tag{3.15}$$

By induction this implies that $Q_1(\mathcal{J}, x^*) \leq \sigma$ under this norm and hence by Theorem 3.4 that

$$\sigma = R_1(\mathcal{J}, x^*) = Q_1(\mathcal{J}, x^*). \tag{3.16}$$

A matrix $A \in L(\mathrm{R}^n)$ is said to be of class M if $\|A\| = \rho(A)$ under some norm. Equivalently, A is of class M if and only if, for any eigenvalue λ with $|\lambda| = \rho(A)$, all Jordan blocks containing λ are one-dimensional (see, e.g., Householder [134, p. 46]). This includes, for example, all diagonalizable matrices, and hence (by Theorem 2.2) covers the case when G is a gradient map.

The condition that $DG(x^*)$ belongs to the class M is not necessary for (3.16) to hold as the following example of Voigt [270] shows:

$$Gx = \begin{pmatrix} x_1^2 + x_2 \\ x_1^4 - x_2^2 \end{pmatrix}, \quad x^* = 0, \tag{3.17}$$

where $DG(x^*)$ is not of class M but (3.16) holds with $\sigma = 0$. This example might suggest that superlinear convergence in the R-measure implies the same in the Q-measure. This is not generally true as the example

$$Gx = \begin{pmatrix} x_1^2 - x_2 \\ x_2^2 \end{pmatrix}, \quad x^* = 0,$$

shows, where $\rho(DG(x^*)) = 0$ but $Q_1(\mathcal{J}, x^*) > 0$. However, for $DG(x^*) = 0$ this cannot happen. In fact, then we can conclude from (3.11) that now (3.15) holds with $\sigma = 0$ and hence that $Q_1(\mathcal{J}, x^*) = 0$. At the same time, note that in the example (3.17) Q-superlinearity occurs even though $DG(x^*) \neq 0$.

Under a stronger differentiability assumption, the condition $DG(x^*) = 0$ actually permits a conclusion about higher-order convergence.

THEOREM 3.6. *Suppose that $G : E \subset \mathrm{R}^n \mapsto \mathrm{R}^n$ has a fixed point $x^* \in \mathrm{int}\,(E)$ and that G is C^1 on an open neighborhood $\mathcal{U} \subset E$ of x^*. If $DG(x^*) = 0$ and DG is F-differentiable at x^*, then x^* is a point of attraction of the process $\mathcal{J}$ defined by (3.9) and $O_R(\mathcal{J}, x^*) \geq O_Q(\mathcal{J}, x^*) \geq 2$. Moreover, if $D^2G(x^*)(h,h) \neq 0$ for all $h \neq 0$ in R^n, then both orders are equal to two.*

Proof. Clearly $x \in \mathcal{U} \mapsto H(x) = DG(x) - DG(x^*) - D^2G(x^*)(x - x^*)$ defines a continuous mapping H on $\mathcal{U}$ and for $\epsilon > 0$ there exist $\delta > 0$ such that

$$\|H(x)\| \leq \epsilon \|x - x^*\|, \quad \forall x \in \mathcal{B} = B(x^*, \delta) \subset E. \tag{3.18}$$

Then, because of $DG(x^*) = 0$, the integral mean value theorem implies that for any $x \in \mathcal{B}$

$$\begin{aligned} \|Gx - x^*\| &= \|\int_0^1 [DG(x^* + t(x - x^*)) - DG(x^*)](x - x^*)dt\| \\ &= \|\int_0^1 [tD^2G(x^*)(x - x^*) + H(x^* + t(x - x^*))](x - x^*)dt\| \\ &\leq \frac{1}{2}(\|D^2G(x^*)\| + \epsilon)\|x - x^*\|^2, \end{aligned} \tag{3.19}$$

which, by Definition 3.2 and Theorem 3.4, proves the first part of the theorem. If the definiteness condition holds for D^2G then there exists a constant $c > 0$ such that $\|D^2G(x^*)(h,h)\| \geq c\|h\|^2$. We may assume that ϵ in (3.18) was chosen such that $2\epsilon \leq c$. Then, by estimating the second line of (3.19) downward, we obtain

$$\|Gx - x^*\| \geq \frac{1}{2}(c-\epsilon)\|x - x^*\|^2 \geq \frac{c}{4}\|x - x^*\|^2, \quad \forall\, x \in \mathcal{B},$$

and hence the result follows from the definition of the Q-factors. □

This theorem extends to higher-order methods. If G is C^p on $\mathcal{U}$, and $D^jG(x^*) = 0$, for $j = 1, 2, \ldots, p$, and if the F-derivative $D^{p+1}G(x^*)$ exists and satisfies $D^{p+1}G(x^*)(h, h, \ldots, h) \neq 0$ for all $h \neq 0$ in R^n, then $O_R(\mathcal{J}, x^*) = O_Q(\mathcal{J}, x^*) = p + 1$. For $n = 1$ this is a classical result of E. Schröder [233].

So far, the results in this section apply only to the simple stationary, one-step iteration (3.9). There are various generalizations to more complicated processes. We consider only the case when the step algorithm $\mathcal{G}$ generates the next iterate x^{k+1} as a solution of an equation of the form

$$G(x^{k+1}, x^k) = 0, \qquad k = 0, 1, \ldots. \tag{3.20}$$

It is easily shown (see [OR] p. 325) that, when the implicit function theorem applies, then Theorem 3.5 can be extended to cover this case.

THEOREM 3.7. *Let $G : E \times D \subset \mathrm{R}^n \times \mathrm{R}^n \mapsto \mathrm{R}^n$ be C^1 on an open neighborhood $\mathcal{U} \subset E$ of a point $x^* \in E$ where $G(x^*, x^*) = 0$. Denote the partial derivatives of G with respect to its two vector variables by $\partial_1 G$ and $\partial_2 G$. If $\partial_1 G(x^*, x^*)$ is nonsingular and $\sigma = \rho(-\partial_1 G(x^*, x^*)^{-1}\partial_2 G(x^*, x^*)) < 1$, then x^* is a point of attraction of the process $\mathcal{J}$ defined by* (3.20), *and $R_1(\mathcal{J}, x^*) = \sigma$.*

This result was extended by Voigt [270] and Cavanagh [49] to a stationary, sequential m-step method of the form

$$x^{k+1} = G(x^k, x^{k-1}, \cdots, x^{k-m+1}), \quad k = m, m+1, \cdots, \tag{3.21}$$

and to other related processes.

3.4 On Efficiency and Accuracy

As indicated before, a specification of the rate of convergence of an iterative process represents only one part in the assessment of its computational complexity. In addition to the orders and convergence factors defined above, we require measures of the effort needed in executing the step algorithm $\mathcal{G}$ of (3.1). Such measures are, of course, strongly dependent on the particular process, and it is therefore understandable that no general definitions are available for them. In defining such measures one has to take into account the sizes of the memory sets M_k as well as the amount of new information computed during each execution of $\mathcal{G}$. Ostrowski [196] proposed the name "horner" for one unit of this computational work. Many authors define a horner as one function call in the step algorithm $\mathcal{G}$ of the process (3.1). Thus, one evaluation of the mapping $F : \mathrm{R}^n \mapsto \mathrm{R}^n$ requires, in general, n horners, while a computation of its derivative may involve as many as n^2 horners.

Suppose that some quantity μ has been established as the maximal (or possibly average) cost of a one-time execution of the step algorithm $\mathcal{G}$ for the particular problem at hand, and let p be the Q- or R-order of $\mathcal{J}$. Traub [265] proposed the quotient p/μ as a measure of the *informational efficiency* of $\mathcal{J}$. Alternatively, Ostrowski [196], and also Traub [265], introduced the *efficiency index* $p^{1/\mu}$. In order to interpret the latter quantity, let p denote the R-order of $\mathcal{J}$. It is easily shown that

$$p = \lim_{k\to\infty} \frac{\log \|x^{k+1} - x^*\|}{\log \|x^k - x^*\|},$$

provided the limit exists. Evidently, $-\log \epsilon_j$ is some measure of the significant digits of $\epsilon_j \in (0, 1)$. Hence, the quotients on the right reflect the proportional gain in significant digits in the approximation during one step. The efficiency index $p^{1/\mu}$, therefore, characterizes the asymptotic value of this gain averaged over the number of horners expended in $\mathcal{G}$.

The two efficiency measures are based only on the order of $\mathcal{J}$. This is adequate for the comparison of processes with order larger than one. Obviously, in the case of processes of order one the convergence factors need to be taken into account. For R-linear processes we may follow here the approach used by Kahan [139] for linear systems (see also Varga [269]). Define

$$\sigma_k = -\frac{1}{\mu k} \log \frac{\|x^k - x^*\|}{\|x^0 - x^*\|}$$

as the average improvement factor for each horner expended during the first k steps, and note that for given $\epsilon > 0$

$$\frac{\|x^k - x^*\|}{\|x^0 - x^*\|}^{1/k} \leq R_1(\mathcal{J}, x^*) + \epsilon \quad \forall\, k \geq k_0(\epsilon).$$

Now introduce $\eta = (-\log R_1)/\mu$ as the linear efficiency index of $\mathcal{J}$. Then

$$\sigma_k \geq \eta \frac{\log(R_1 + \epsilon)}{\log R_1}, \quad \forall\, k \geq k_0$$

characterizes η as an asymptotic lower bound for the average improvement factors.

Once efficiency measures for a class of processes have been established, an important question is the determination of optimal members of this class. As mentioned before, Traub, Woźniakowski, and their coworkers have developed a substantial theory addressing this question; principal references are the two monographs by Traub and Woźniakowski [267] and Traub, Wasilkowski, and Woźniakowski [266]. Other work in this area includes Renegar [209], Shub and

Smale [246], and Smale [249]. These interesting theories are beyond our scope here.

Generally, the various efficiency measures do not reflect an important computational aspect of any iterative process, namely, its stability. This concerns questions about the sensitivity of a process to influences such as perturbation of the mapping F, an increase in the dimension of the problem, variation of the input of $\mathcal{J}$, and, last but not least, the use of finite state arithmetic. Various examples are available which indicate that the effect of such influences upon an iterative process ought not to be disregarded. We comment here only briefly on the possible impact of floating point arithmetic. Roundoff errors are expected in the evaluation of the function (or, where needed, its derivatives), as well as in the execution of the step routine itself. In general, these errors may result in a slowdown of the convergence rate or even a breakdown of the entire process. In addition, it is important to observe that the roundoff errors in the function evaluation introduce an uncertainty region for the function values. As an illustrative example we consider the quartic polynomial

$$\begin{aligned} &p(x) = a_0x^4 + a_1x^3 + a_2x^2 + a_1x + a_4, \\ &a_0 = 1, \ a_1 = -202, \ a_2 = 15299, \ a_3 = -514898, \ a_4 = 6497400, \end{aligned}$$

which has the zeros $x^* = 49, 50, 51, 52$. In this case it is natural to evaluate $p(x)$ by means of the standard Horner scheme:

(3.22)

input: x
$p := a_0$;
for $k = 1, \ldots, 4$ **do** $p := p * x + a_k$;
output: p

For the computation the floating-point processor of a Pentium-Pro chip was used. For points in a neighborhood of the roots, Table 3.1 shows the absolute error between $p(x)$, computed by (3.22) in single precision, and the value $(x-49)(x-50)(x-51)(x-52)$, computed in double precision. For any x in the domain D of the function F and a specific floating-point evaluation $\mathrm{fl}(F(x))$ of F, we define $\|\mathrm{fl}(F(x)) - F(x)\|$ as the radius of the uncertainty ball at x. For a subset $D_0 \subset D$ the maximum uncertainty radius at the points of D_0 is the uncertainty radius of F on that set. The ideal situation would be, of course, when this uncertainty radius is a modest multiple of the unit roundoff of the particular floating-point calculation. This is typically expected of any library function, such as $\sin, \cos$, etc.

But in our example the uncertainty radius of the quartic polynomial on the set of x-values in the table is 1.6438. This is mainly due to severe subtractive cancelling in the evaluation of the last term $p * x + a_4$ of (3.22). In double-precision arithmetic the uncertainty radius becomes 0.1036(−08) which is much better but still considerably larger than the hoped-for small multiple of the unit roundoff. Clearly, with errors in the function evaluation as large as those in our single-precision calculation there is little hope that an iterative process can reliably find any of the roots. The example certainly shows that, in the solution

TABLE 3.1
Errors in polynomial evaluation.

x	error	x	error	x	error
47.0	0.0000	49.0	0.0000	51.0	0.0000
47.2	0.7788(–01)	49.2	–0.4944	51.2	–0.8744
47.4	–1.1394	49.4	–0.5835	51.4	–0.3021
47.6	0.5177	49.6	0.5688	51.6	0.5437
47.8	0.2471	49.8	0.7469	51.8	0.2349
48.0	0.0000	50.0	0.0000	52.0	0.0000
48.2	–0.2776	50.2	–0.5631	52.2	0.1889
48.4	–1.3967	50.4	–1.3906	52.4	–0.8901
48.6	1.6438	50.6	1.1099	52.6	1.2701
48.8	0.9495	50.8	–0.1989(–01)	52.8	0.3348

of any problem, it is important to assess the accuracy of the function evaluations and to ensure that the uncertainty radii remain as small as possible.

CHAPTER 4

Methods of Newton Type

This chapter introduces several basic types of iterative methods derived by linearizations, including, in particular, the classical Newton method and a few of its modifications. Throughout the chapter, $F : E \subset \mathrm{R}^n \mapsto \mathrm{R}^n$ denotes a mapping defined on some open subset E of R^n.

4.1 The Linearization Concept

Let $x^k \in E$ be a current iterate when solving $Fx = 0$. The idea of a *linearization method* is to construct an affine approximation

$$L_k : \mathrm{R}^n \mapsto \mathrm{R}^n, \quad L_k x = A_k(x - x^k) + Fx^k, \quad A_k \in L(\mathrm{R}^n), \tag{4.1}$$

that agrees with F at x^k, and to use a solution of $L_k x = 0$ as the next iterate. We shall not consider here the situation when some or all of the matrices are allowed to be singular, and assume always that all A_k, $k \geq 0$ in (4.1) are invertible. Then the resulting (nonsingular) linearization method becomes

$$x^{k+1} = x^k - A_k^{-1} Fx^k, \qquad k = 0, 1, \ldots. \tag{4.2}$$

In terms of the iterative algorithm (3.1) this means that the step algorithm $\mathcal{G}$ now has the following form.

(4.3)

$\mathcal{G}$: **input:** $\{k, x^k, M_k\}$
evaluate Fx^k; construct the matrix A_k;
solve $A_k y = Fx^k$ for y;
if solver failed **then return:** fail;
$x^{k+1} = x^k - y$;
return: $\{x^{k+1}, M_{k+1}\}$

As in the case of iterative processes for linear systems it may be desirable to introduce a step relaxation (damping); that is, to lengthen or shorten the step y between the two iterates x^k and x^{k+1} by some factor $\omega_k > 0$. This means that (4.2) is changed to

$$x^{k+1} = x^k - \omega_k A_k^{-1} Fx^k, \qquad k = 0, 1, \ldots \tag{4.4}$$

or, simply, that the next iterate x^{k+1} in $\mathcal{G}$ is $x^{k+1} = x^k - \omega_k y$. In order to enforce the nonsingularity of the A_k, it is often useful to replace A_k by $A_k + \lambda_k I$ where λ_k is a suitable shift parameter. This changes (4.2) to

$$x^{k+1} = x^k - (A_k + \lambda_k I)^{-1} F x^k, \qquad k = 0, 1, \ldots, \tag{4.5}$$

and may be called a *regularization* of (4.2). Any step reduction or regularization will be reliably efficient only if it is based on adaptively obtained information about the performance of the iteration. Examples for this will be discussed in section 6.4 and chapter 8.

The simplest linearization methods are the *(parallel) chord methods* where all matrices A_k are identical. In other words, chord methods have the form

$$x^{k+1} = x^k - A^{-1} F x^k, \qquad k = 0, 1, \ldots, \tag{4.6}$$

with some constant, invertible matrix $A \in L(\mathrm{R}^n)$. Typical examples include the Picard iteration with $A = \alpha I$, $\alpha \neq 0$, and the chord Newton method with $A = DF(x^0)$. A special case of the Picard iteration arises if F has the form $Fx = Bx - Gx$ with nonsingular $B \in L(\mathrm{R}^n)$ and a nonlinear mapping $G : E \subset \mathrm{R}^n \mapsto \mathrm{R}^n$. Here the chord method (4.6) with the choice $A = B$ becomes

$$x^{k+1} = A^{-1} G x^k, \quad k = 0, 1, \ldots,$$

which, for $B = I$; that is, for the fixed-point equation $x = Gx$, is the classical *method of successive approximations*

$$x^{k+1} = G x^k, \quad k = 0, 1, \ldots. \tag{4.7}$$

The local convergence of any chord method (4.6) follows immediately from Theorem 3.5. In fact, if F is F-differentiable at x^*, then the iteration mapping $\Phi x = x - A^{-1} F x$ corresponding to (4.6) has at x^* the F-derivative $\Phi'(x^*) = I - A^{-1} DF(x^*)$. Hence, if $\sigma = \rho(\Phi'(x^*)) < 1$, then x^* is a point of attraction of the process $\mathcal{J}$ defined by (4.6) and $R_1(\mathcal{J}, x^*) = \sigma$.

At the same time, for the convergence of the simple methods of the form (4.7) we also have the contraction-mapping theorem, which in our setting may be phrased as follows.

THEOREM 4.1. *Suppose that $G : E \subset \mathrm{R}^n \mapsto \mathrm{R}^n$ maps a closed set $C \subset E$ into itself and that there exists a constant $\alpha \in (0, 1)$ such that*

$$\|Gy - Gx\| \leq \alpha \|y - x\|, \quad \forall\, x, y \in C. \tag{4.8}$$

Then G has a unique fixed point $x^* = Gx^*$ in C and for any $x^0 \in C$ the iterates (4.7) converge to x^* and satisfy

$$\|x^k - x^*\| \leq \frac{\alpha^k}{1 - \alpha} \|x^1 - x^0\|, \quad k \geq 0. \tag{4.9}$$

Proof. Any two fixed points $x^*, y^* \in C$ satisfy

$$\|y^* - x^*\| = \|Gy^* - Gx^*\| \le \alpha \|y^* - x^*\|,$$

which, because $\alpha < 1$, implies that $y^* = x^*$. Since $GC \subset C$, the sequence $\{x^k\}$ of (4.7) is well defined and remains in C. For any $k, m > 0$ we obtain

$$\begin{aligned}
\|x^{k+m+1} - x^k\| &\le \sum_{j=0}^{m} \|x^{h+j+1} - x^{k+j}\| \\
&\le \sum_{j=0}^{m} \alpha^j \|x^{k+1} - x^k\| \le \frac{1}{1-\alpha}\|x^{k+1} - x^k\| \qquad (4.10)\\
&\le \frac{\alpha^k}{1-\alpha}\|x^1 - x^0\|,
\end{aligned}$$

which shows that $\{x^k\}$ is a Cauchy sequence. Thus $\lim_{k\to\infty} x^k = x^*$ exists and the closedness of C implies that $x^* \in C$. From (4.9) it follows that G is continuous whence (4.7) gives $x^* = Gx^*$. The error estimate (4.9) is a consequence of (4.10) for $m \to \infty$. □

4.2 Methods of Newton Form

The definition of linearization methods in the previous section does not include a mechanism for constructing the matrices A_k in (4.2). A simple selection principle involves a given matrix-valued mapping $A : E \subset \mathrm{R}^n \mapsto L(\mathrm{R}^n)$ which is used to define the iteration matrices at the iterates x^k by $A_k = A(x^k)$. In other words, these methods have the form

$$x^{k+1} = x^k - A(x^k)^{-1}Fx^k, \qquad k = 0, 1, \ldots. \qquad (4.11)$$

The most famous member of this class is Newton's method

$$\mathcal{N}: \quad x^{k+1} = x^k - DF(x^k)^{-1}Fx^k, \quad k = 0, 1, \ldots, \qquad (4.12)$$

in which the affine approximation L_k of (4.1) is obtained by truncating the Taylor expansion of F at x^k after the linear term. Accordingly, the linearization processes (4.11) are also called *methods of Newton form.*

Most local convergence results for methods of Newton form assume that the desired solution is a simple zero in the following sense.

DEFINITION 4.1. *A zero $x^* \in E$ of F is simple if F is F differentiable at x^* and $DF(x^*)$ is nonsingular.*

Evidently, in the scalar case this reduces exactly to the standard definition of simple roots of differentiable functions and, as in that case, Definition 4.1 implies that simple zeros are locally unique.

THEOREM 4.2. *Let $x^* \in E$ be a simple zero of F, and set $\beta = \|DF(x^*)\|$, $\gamma = \|DF(x^*)^{-1}\|$. Then, given $0 < \epsilon < 1/\gamma$, there exists a closed ball $\mathcal{B} = \bar{B}(x^*, \delta) \subset E$, $\delta > 0$, such that*

$$(\beta + \epsilon)\|x - x^*\| \ge \|Fx\| \ge \left(\frac{1}{\gamma} - \epsilon\right)\|x - x^*\|. \qquad (4.13)$$

In particular, x^* is the only zero of F in $\mathcal{B}$.

Proof. For any ϵ such that $0 < \epsilon < 1/\gamma$, choose $\delta > 0$ such that $\mathcal{B} = \bar{B}(x^*, \delta) \subset E$ and

$$\|Fx - Fx^* - DF(x^*)(x - x^*)\| \leq \epsilon\|x - x^*\|, \quad \forall x \in \mathcal{B}.$$

Then, by applying the triangle inequality to

$$Fx = Fx - Fx^* - DF(x^*)(x - x^*) + DF(x^*)(x - x^*),$$

we obtain (4.13), which proves the result. □

Under the conditions of Theorem 4.2, for any sequence $\{x^k\} \subset E$ converging to $x^* \in E$ and satisfying $x^k \neq x^*$ for $k \geq k_0$, we obtain

$$\left(\frac{1}{\gamma} - \epsilon\right)\frac{\|x^{k+1} - x^*\|}{\|x^k - x^*\|} \leq \frac{\|Fx^{k+1}\|}{\|x^k - x^*\|} \leq (\beta + \epsilon)\frac{\|x^{k+1} - x^*\|}{\|x^k - x^*\|}, \ \forall x^k \in \mathcal{B},$$

whence the convergence is Q-superlinear if and only if

$$\lim_{k\to\infty} \frac{\|Fx^{k+1}\|}{\|x^k - x^*\|} = 0, \quad x^k \neq x^* \text{ for } k \geq k_0. \tag{4.14}$$

The following result provides the basis for the local convergence analysis of methods of Newton form.

THEOREM 4.3. *Let F be F-differentiable at the zero $x^* \in E$ of F, and suppose that the mapping $A : \mathcal{U} \mapsto L(\mathrm{R}^n)$ is defined on some open neighborhood $\mathcal{U} \subset E$ of x^* and is continuous at x^* with nonsingular $A(x^*)$. Then there is a ball $\mathcal{B} = \bar{B}(x^*, \delta) \subset \mathcal{U}$, $\delta > 0$, on which the mapping $x \in \mathcal{B} \mapsto Gx = x - A(x)^{-1}Fx \in \mathrm{R}^n$, is well defined and G has at x^* the F-derivative $DG(x^*) = I - A(x^*)^{-1}DF(x^*)$.*

Proof. Set $\beta = \|A(x^*)^{-1}\|$ and for given $\epsilon > 0$, $2\beta\epsilon < 1$, choose $\delta > 0$ such that $\mathcal{B} = \bar{B}(x^*, \delta) \subset \mathcal{U}$ and $\|A(x) - A(x^*)\| \leq \epsilon$, $\forall x \in \mathcal{B}$. Then $A(x)$ is invertible for all $x \in \mathcal{B}$, and

$$\|A(x)^{-1}\| \leq \frac{\beta}{1 - \beta\epsilon} < 2\beta, \quad \forall x \in \mathcal{B}; \tag{4.15}$$

that is, the mapping G is well defined on $\mathcal{B}$. Since F is F-differentiable at x^* we may choose $\delta > 0$ small enough such that

$$\|Fx - Fx^* - DF(x^*)(x - x^*)\| \leq \epsilon\|x - x^*\|, \quad \forall x \in \mathcal{B}.$$

Then the F-differentiability of G at x^* follows from the estimates

$$\begin{aligned}
&\|Gx - Gx^* - \left(I_n - A(x^*)^{-1}DF(x^*)\right)(x - x^*)\| \\
&\quad = \|A(x^*)^{-1}\, DF(x^*)(x - x^*) - A(x)^{-1}Fx\| \\
&\quad \leq \|- A(x)^{-1}\,[Fx - Fx^* - DF(x^*)(x - x^*)]\,\| \\
&\qquad + \|A(x)^{-1}\,[A(x^*) - A(x)]\,A(x^*)^{-1}DF(x^*)(x - x^*)\| \\
&\quad \leq \epsilon\left[\,2\beta + 2\beta^2\,\|DF(x^*)\|\,\right]\,\|x - x^*\|,
\end{aligned} \tag{4.16}$$

since $\epsilon > 0$ is arbitrarily small and the term in square brackets on the last line of (4.16) is constant. ☐

Under the conditions of Theorem 4.3, we have for a process $\mathcal{J}$ of Newton form (4.11) the local convergence condition

$$\sigma = \rho(I - A(x^*)^{-1}DF(x^*)) < 1, \tag{4.17}$$

which implies that $R_1(\mathcal{J}, x^*) = \sigma$. In particular, for Newton's method we obtain the following attraction theorem.

THEOREM 4.4. *Let F be C^1 on an open neighborhood $\mathcal{U} \subset E$ of its simple zero $x^* \in E$. Then* (i) *x^* is a point of attraction of the Newton method $\mathcal{N}$ of* (4.12) *and $R_1(\mathcal{N}, x^*) = Q_1(\mathcal{N}, x^*) = 0$;* (ii) *there exists a ball $\mathcal{B} = \bar{B}(x^*, \delta) \subset \mathcal{U}$, $\delta > 0$, such that $x^k \in \mathcal{B}$ and $\|Fx^{k+1}\| \le \|Fx^k\|$ for all sufficiently large k;* (iii) *if the additional condition*

$$\|DF(x) - DF(x^*)\| \le \alpha \|x - x^*\|^q, \qquad \forall\, x \in \mathcal{U} \tag{4.18}$$

holds with $0 < \alpha < \infty$ and $q \ge 0$, then $O_R(\mathcal{N}, x^*) \ge O_Q(\mathcal{N}, x^*) \ge 1 + q$.

Proof. (i) With A specified by $x \in \mathcal{U} \mapsto A(x) = DF(x)$, it follows from Theorem 4.3 that the Newton function $Gx = x - DF(x)^{-1}Fx$ is defined on some ball $\mathcal{B} = \bar{B}(x^*, \delta) \subset \mathcal{U}$, $\delta > 0$, and that $DG(x^*) = 0$. Then Theorem 3.6 ensures that $R_1(\mathcal{N}, x^*) = Q_1(\mathcal{N}, x^*) = 0$.

(ii) From (i) it follows that $x^k \in \mathcal{B}$ for $k \ge k_0$ and sufficiently large k_0. Set $\gamma = \|DF(x^*)\|$. By shrinking δ if needed we have $\|DF(x) - DF(x^*)\| \le 1/(3\gamma)$ for $x \in \mathcal{B}$ whence the perturbation lemma ensures the invertibility of $DF(x)$ and $\|DF(x)^{-1}\| \le 3\gamma/2$ for $x \in \mathcal{B}$. Therefore

$$\begin{aligned} \|Fx^{k+1}\| &= \|Fx^{k+1} - Fx^k - DF(x^k)(x^{k+1} - x^k)\| \\ &= \left\| \int_0^1 [DF(x^k + t(x^{k+1} - x^k)) - DF(x^k)](x^{k+1} - x^k)dt \right\| \\ &\le \frac{2}{3\gamma}\|x^{k+1} - x^k\| = \frac{2}{3\gamma}\|DF(x^k)^{-1}Fx^k\| \le \|Fx^k\|. \end{aligned}$$

(iii) If (4.18) holds then we have

$$\|Fx - Fx^* - DF(x^*)(x - x^*)\| \le \alpha \|x - x^*\|^{q+1}, \quad \forall\, x \in \mathcal{U},$$

whence, as in (4.16), we obtain, for $x \in \mathcal{B}$,

$$\begin{aligned} \|Gx - x^*\| &= \| - DF(x)^{-1}[Fx - Fx^* - DF(x)(x - x^*)]\| \\ &\le \frac{3\gamma\alpha}{2}\|x - x^*\|^{q+1}, \end{aligned}$$

which implies that $O_R \ge O_Q \ge 1 + q$. ☐

A closer inspection of the proof shows the validity of part (i) under the weaker assumption that F is G-differentiable on $\mathcal{U} \subset E$ and DF is continuous at x^* (whence F is again F-differentiable at x^*).

The theorem shows that when $q = 1$ in (4.18); that is, when DF is Lipschitz continuous at x^*, then Newton's method converges at least Q-quadratically. As in the proof of Theorem 3.6 we can show that when the second F-derivative of F at x^* exists and satisfies $D^2F(x)(h,h) \neq 0$ for all $h \neq 0$ in R^n, then (4.18) holds with $q = 1$ and $O_R(\mathcal{N}, x^*) = O_Q(\mathcal{N}, x^*) = 2$ (see section 10.2 of [OR]).

4.3 Discretized Newton Methods

The execution of a step of any linearization method (4.3) requires the
(a) evaluation of the n components of Fx^k,
(b) evaluation of the n^2 elements of A_k,
(c) numerical solution of the $n \times n$ linear system,
(d) work needed to handle the memory set M_{k+1}.

Evidently, all these tasks depend not only on the particular method but also on the specific problem at hand. For instance, for high-dimensional systems, sparse-matrix techniques are required for (c) and, instead of direct solvers, iterative methods may be needed. In the case of Newton's method, (b) involves the evaluation of the n^2 first partial derivatives of F. Algebraic expressions for these partial derivatives are not always easily derivable, in which case some automatic differentiation method may become essential (see, e.g., Griewank and Corliss [119] where an extensive bibliography may be found). Another approach is the approximation of the partial derivatives by suitable finite-difference approximations. This leads to the so-called discrete Newton methods.

For a point x in the domain of F and suitable parameter vector $h \in \mathrm{R}^m$ let $J(x,h) \in L(\mathrm{R}^n)$ be a matrix whose elements $J(x,h)_{ij}$ are difference approximations of the partial derivatives $\partial f_i(x)/\partial x_j$. Then—provided it is well defined—we call the linearization process

$$(4.19) \qquad x^{k+1} = x^k - J(x^k,h^k)^{-1}Fx^k, \qquad k = 0,1,\ldots, \; h^k \in \mathrm{R}^m,$$

a *discretized Newton method.* A simple example for the matrix $J(x,h)$ is, of course,

$$(4.20) \qquad J(x,h) \in L(\mathrm{R}^n), \; J(x,h)_{ij} = \frac{1}{h_{i,j}}(f_i(x + h_{i,j}e^j) - f_i(x)),$$

where, as usual $e^1,\ldots,e^n$ are the natural basis vectors of R^n and $h \in \mathrm{R}^{n^2}$ is a vector with small nonzero components. Evidently, this is a special case of the more general discretization

$$(4.21) \quad \left\{ \begin{array}{l} J(x,h) \in L(\mathrm{R}^n), \; J(x,h)_{ij} = \dfrac{1}{h_{ij}}\Big[f_i(x + \beta\displaystyle\sum_{\ell=1}^{j-1} h_{i\ell}e^\ell + h_{ij}e^j) \\ \qquad\qquad\qquad\qquad\qquad\qquad -f_i(x + \beta\sum_{\ell=1}^{j-1} h_{i\ell}e^\ell)\Big]. \end{array} \right.$$

For the convergence study of the processes (4.19) we need to know how $J(x,h)$ approximates $DF(x)$. Clearly, the iterative process (4.19) works only with specific (and typically nonzero) vectors h^k and, minimally, we should expect that for each $k \geq 0$ the matrix $J(x^k, h^k)$ is sufficiently close to $DF(x^k)$. Since the form of J is generally fixed and only the vectors h^k remain to be chosen, this approximation condition is usually ensured by assuming that $J(x^k, h)$ converges to $DF(x^k)$ when h tends to zero. Certainly, for the examples (4.20) and (4.21) this follows readily by standard arguments of analysis (see also Theorem 5.2).

These approximation properties may be used in various forms to obtain some local convergence results for discretized Newton methods. We shall postpone the discussion of such results to chapters 5 and 6. Here we address only some questions relating to the efficient computation of certain discretized forms of the Jacobian.

For a given choice of positive steps $h_{i,j}$ the evaluation of the matrix (4.20) at a point x requires the function values $f_i(x)$ and $f_i(x+h_{i,j}e^j)$, for $i, j = 1, \ldots, n$, which, in the terminology of section 3.4, means that we need $n + n^2$ horners. For (4.21) the additional values

$$f_i\left(x + \beta \sum_{k=1}^{j} h_{i,k}e^k\right), \quad i = 1, \ldots, n, \ j = 1, \ldots, n-1,$$

are also used and therefore a total of $2n^2$ horners are needed. On the other hand, for (4.21) with $\beta = 1$ we require only the function values

$$f_i(x), \ f_i\left(x + \sum_{k=1}^{j} h_{i,k}e^k\right), \quad i = 1, \ldots, n, \ j = 1, \ldots, n,$$

and thus, once again, $n + n^2$ horners.

These examples already show that not only the approximation properties of J but also its specific form will have an effect on the overall behavior of the discretized Newton method.

The count in terms of horners is often unrealistic. In fact, in many problems we only have a routine for evaluating the entire vector $Fx \in \mathbb{R}^n$ at a given point x, and the cost of computing one component value $f_i(x)$ is almost as high as that of computing all of Fx. In that case, the above examples are extraordinarily costly unless we restrict the choice of the steps $h_{i,j}$. For instance, if, say, $h_{i,j} \equiv h_j$ for $i = 1, \ldots, n$, then, in the case of (4.20), the entire jth column of $J(x, h)$ can be computed by the single difference

$$\frac{1}{h_j}\left(F(x + h_j e^j) - Fx\right); \tag{4.22}$$

that is, we need altogether $n + 1$ calls to the routine for F. Similarly, for (4.21) with $\beta = 1$ we can compute $J(x, h)$ columnwise by the algorithm

input: $\{ x, h_1, \ldots, h_n \}$
$y := x$; Evaluate $u = Fy$;
for $k = 1, \ldots, n$
  $y := y + h_k e^k$; Evaluate $v := Fy$;
  $q^k := (u - v)/h_k$; $u := v$;
endfor
return: $J(x, h) := (q^1, \ldots, q^n)$

Hence again $n + 1$ evaluations of F are needed.

For sparse Jacobians $DF(x)$ there may be considerable savings in the evaluation of $J(x,h)$. Let $T_F \equiv (\tau_{i,j}) \in L(\mathrm{R}^n)$ be the pattern matrix of F (see Definition 2.1). We introduce a partition $N = N_1 \cup N_2 \cup \cdots N_m$ of the index set $N = \{1,2,\ldots,n\}$ into mutually disjoint subsets N_k with the property that for every $i \in N$ there exists in each N_k at most one $j \in N_k$ with $\tau_{i,j} = 1$. For example, suppose that $DF(x)$ has a tridiagonal pattern matrix. Then, there exists a partition into three sets $N = N_1 \cup N_2 \cup N_3$ with $N_k = \{i \in N \ : \ i = k + 3(j-1), \ j = 1,2,\ldots\}$. For instance, for $n = 10$ an example of such a partition is

$$\{1,2,\ldots,10\} = \{1,4,7,10\} \cup \{2,5,8\} \cup \{3,6,9\}. \tag{4.23}$$

Clearly the choice of such a partition is not unique.

Consider now the computation of the matrix $J(x,h)$ with the columns (4.22). Let $N_k = \{j_1,\ldots,j_p\}$ be one of the subsets of a partition and calculate the difference vector

$$d^k = F(x + h_{j_1}e^{j_1} + h_{j_2}e^{j_2} + \cdots + h_{j_p}e^{j_p}) - Fx.$$

For instance, for the set $N_1 = \{1,4,7,10\}$ of the partition (4.23) we get the vector

$$d^1 = \begin{pmatrix} f_1(x + h_1e^1) & - f_1(x) \\ f_2(x + h_1e^1) & - f_2(x) \\ f_3(x + h_4e^4) & - f_3(x) \\ f_4(x + h_4e^4) & - f_4(x) \\ f_5(x + h_4e^4) & - f_5(x) \\ f_6(x + h_7e^7) & - f_6(x) \\ f_7(x + h_7e^7) & - f_7(x) \\ f_8(x + h_7e^7) & - f_8(x) \\ f_9(x + h_{10}e^{10}) & - f_9(x) \\ f_{10}(x + h_{10}e^{10}) & - f_{10}(x) \end{pmatrix}$$

By construction of the partition, for each $i \in N$ there is at most one $j_i \in N_k$ such that $\tau_{i,j_i} = 1$. Then, the ith component $(e^i)^\top d^k$ of the vector d^k divided by the step h_{j_i} is exactly the (i,j_i)th element of $J(x,h)$.

This shows that, for a column partition of m subsets, we require only $m + 1$ evaluations of F to compute the matrix $J(x,h)$. In our tridiagonal example this means that $J(x,h)$ can be generated with four evaluations of F for any dimension $n \geq 3$. For a general problem involving sparse Jacobians, the construction of the required partition, of course, needs to be done only once at the start of the computations.

The approach was first proposed by Curtis, Powell, and Reid [58] and further analyzed by Coleman and Moré [53], and Coleman, Garbow, and Moré [52]. In [53] it is shown that the construction of the partition is closely related to graph-coloring algorithms.

4.4 Attraction Basins

All the iterative processes considered in the previous two sections had the form

$$x^{k+1} = Gx^k, \quad k = 0, 1, \dots \tag{4.24}$$

and our interest centered on their convergence to a fixed point $x^* = Gx^*$ which then turned out to be a zero of the given mapping F. The iteration (4.24) defines a discrete dynamical process. For ease of notation suppose that the mapping G is defined on all of R^n. The set of all starting points $x^0 \in \mathrm{R}^n$ for which (4.24) converges to the fixed point x^* is called the *attraction basin* of the process at x^*.

The attraction theorems discussed so far are local in nature and ensured, for a specific zero x^* of F, only the existence of an open neighborhood of x^* contained in the attraction basin of G at x^*. It is important to note that, in general, the boundaries between attraction basins for different zeros have a complicated structure and, in fact, are so-called *fractal boundaries*. This is a property of discrete as well as continuous dynamical systems. The literature in this area has been growing rapidly, and the topic is outside our scope. We sketch only a few ideas following, in part, Barnsley [20].

For ease of notation it is useful to extend our setting and define, more generally, a discrete dynamical system as a pair (X, G) consisting of a mapping $G : X \mapsto X$ from a metric space X into itself. The (forward) iterates of G are the mappings $G^{\circ k} : X \mapsto X$ defined recursively by $G^{\circ 0}x = x$, $G^{\circ 1}x = Gx$, and $G^{\circ k+1}x = G^{\circ k}x$, for $k = 0, 1, \dots$ and all $x \in X$. The *orbit* of a point $x \in X$ is the sequence $\{G^{\circ k}x\}_{k=0}^{\infty}$. A *periodic point* of G is a point $x \in X$ such that $G^{\circ k}x = x$ for some integer $k \geq 1$ called a period of x. In other words, a periodic point x of G with period k is a fixed point of the mapping $G^{\circ k}$. A periodic point $x^* \in X$ of G with period k is *repulsive* if there exist constants $\delta > 0$ and $c > 0$ such that

$$\rho(G^{\circ k}x, G^{\circ k}x^*) \geq c\rho(x, x^*), \quad \forall\, x \in X,\ \rho(x, x^*) \leq \delta,$$

where $\rho : X \times X \mapsto \mathrm{R}^1$ denotes the metric on X.

The study of the complexity of the orbit structure of rational functions goes back, at least, to Julia [137] and Fatou [92]. Let $\bar{\mathrm{C}}$ denote the Riemann sphere; that is, the complex plane C together with the point at infinity. The *Julia set* of a rational function $G : \bar{\mathrm{C}} \mapsto \bar{\mathrm{C}}$ of degree at least two is defined as the closure of the set of repulsive periodic points of G. Suppose that G is a polynomial of degree greater than one and consider the set

$$K_F = \left\{ z \in \mathrm{C} \ : \ |G^{\circ k}x| \leq c, \forall\, k \geq 0, \ \text{for some constant } c > 0 \right\}$$

of all points of C whose orbits do not converge to the point at infinity. Then the boundary J_G of K_G is the Julia set of the polynomial and it can be shown that both J_G and K_G are nonempty, compact subsets of C. Moreover, both sets are invariant in the sense that $G(J_G) = J_G$ and $G(K_G) = K_G$. The complement $\bar{\mathrm{C}} \setminus K_G$ is a nonempty, open, path-connected subset of $\bar{\mathrm{C}}$, the attraction basin of the point at infinity. Examples of K_G for various simple polynomials are shown in Barnsley [20], and interesting graphics involving Julia sets for Newton's method may be found in Curry, Garnett, and Sullivan [57], and Peitgen and Richter [201], and, for some higher order processes, in Vrscay [271].

The "strange" structure of attraction basins has been studied especially for differential-dynamical systems. For instance, for certain planar systems it has been shown that there exist basins where every point on the common boundary between that basin and another basin is also on the boundary of a third basin, (see, e.g., Nusse and Yorke [188], [189]).

These results certainly suggest that, when an iterative process is used to compute a particular zero x^* of a given mapping on R^n, we may well encounter very strange convergence behavior unless the process is started sufficiently near x^*. This represents some justification for our emphasis on local convergence theorems, but also raises various questions. In particular, there is interest in techniques for extending the guaranteed convergence neighborhood. Some answers for this may be found in section 6.4. Another question concerns the estimation of the radii of balls that are contained in the attraction basin of a given zero. For Newton's method, the following example of such a result was given by Den Heijer and Rheinboldt [63].

THEOREM 4.5. *Let $F : E \subset \mathrm{R}^n \mapsto \mathrm{R}^n$ be C^1 and*

$$\|DF(y) - DF(x)\| \leq \gamma \|y - x\|, \quad \forall\, x, y \in E.$$

Then for any simple zero $x^* \in \mathrm{int}\ (E)$ of F any open ball $B(x^*, \rho) \subset E$ with

$$\rho \leq \frac{2}{3\gamma \|DF(x^*)^{-1}\|}$$

is an attraction ball of Newton's method.

The proof is similar to those of Theorems 4.3 and 4.4. Another result of this type is the following theorem of Deuflhard and Potra [74] which uses a form of an *affine-invariant Lipschitz condition* pioneered by Deuflhard (see, e.g., Deuflhard and Heindl [73]).

THEOREM 4.6. *Let $F : E \subset \mathrm{R}^n \mapsto \mathrm{R}^n$ be C^1 on the open, convex set E and suppose that $DF(x)$ is invertible for any $x \in E$ and that, for some constant $\gamma \geq 0$,*

$$\|DF(x)^{-1}[DF(x + sv) - DF(x)]\| \leq s\gamma \|v\|, \quad \forall\, x, x + v \in E, \ s \in [0, 1].$$

Then, for any zero $x^* \in E$ of F, any ball $B(x^*, \rho) \subset E$ with $\rho < 2/\gamma$ is an attraction ball of Newton's method.

CHAPTER 5

Methods of Secant Type

5.1 General Secant Methods

This section continues the discussion of discretized Newton methods begun in section 4.3. The material follows the presentation of [OR].

For a one-dimensional mapping $f : \mathrm{R}^1 \mapsto \mathrm{R}^1$, the method (4.19) with J specified by (4.20) constitutes a linearization method defined by the affine function

$$L_k x = \frac{1}{h^k}\Big[f(x^k + h^k) - f(x^k)\Big](x - x^k) + f(x^k).$$

Evidently, $x \mapsto L_k x$ may be interpreted alternatively as an approximation of the tangent function $x \mapsto Df(x^*)(x - x^k) + f(x^k)$ or as a linear interpolating function which agrees with f at x^k and $x^k + h^k$. The definition (4.19) of the discretized Newton methods represents the first interpretation. For an extension of the second viewpoint to higher dimensions, note that in R^n an affine function

$$L_k x = A_k x + a^k, \quad A_k \in L(\mathrm{R}^n), \quad a^k \in \mathrm{R}^n, \tag{5.1}$$

is determined by its values at $n+1$ interpolating points $x^{k,0}, \ldots, x^{k,n}$, provided these points are in general position; that is, the differences $x^{k,j} - x^{k,0}$, $j = 1, \ldots n$, span R^n. More specifically, the following interpolation result is well known (see, e.g., p. 192 of [OR]).

THEOREM 5.1. *For given $x^{k,j}$, $y^{k,j} \in \mathrm{R}^n$, $j = 0, 1, \ldots, n$, there exists a unique function (5.1) such that $L_k x^{k,j} = y^{k,j}$, $j = 0, 1, \ldots, n$, if and only if the $\{x^{k,j}\}$ are in general position. Moreover, A_k is nonsingular if and only if the $\{y^{k,j}\}$ are in general position.*

In order to apply this to the construction of a linearization process, we have to select, at each iteration step, $n+1$ interpolating points $x^{k,0}, \ldots, x^{k,n} \in E$. For convenience, we will always set $x^{k,0} = x^k$ in which case the points will be in general position if and only if the matrix

$$H_k = (x^{k,1} - x^k, \ldots, x^{k,n} - x^k) \tag{5.2}$$

is invertible. Conversely, (5.2) is equivalent with

$$x^{k,j} = x^k + H_k e^j, \qquad j = 1, \ldots, n, \tag{5.3}$$

and hence, if $H_k \in L(\mathrm{R}^n)$ is any invertible matrix, then (5.3) defines n points which, together with $x^{k,0} = x^k$, are in general position. Thus, it suffices to specify the nonsingular matrix H_k.

If the points $\{x^{k,j}\}$ are in general position, then the affine interpolation function L_k of (5.1) is uniquely determined by the conditions $A_k x^{k,j} + a^k = Fx^{k,j}$, $j = 0, 1, \ldots, n$, and, by taking differences, it follows immediately that

$$A_k = (Fx^{k,1} - Fx^k, \ldots, Fx^{k,n} - Fx^k)H_k^{-1}. \tag{5.4}$$

This shows, indeed, that A_k is nonsingular exactly if the $\{Fx^{k,j}\}$ are in general position.

Since the selection of the points $\{x^{k,j}\}$ is equivalent with the choice of a nonsingular matrix $H = H_k \in L(\mathrm{R}^n)$, we define, as in (5.4),

$$J(x, H) = \Gamma H^{-1}, \; \Gamma = (F(x + He^1) - Fx, \ldots, F(x + He^n) - Fx). \tag{5.5}$$

Then the resulting nonsingular linearization process (4.2) has the form

$$x^{k+1} = x^k - J(x^k, H_k)^{-1} Fx^k, \qquad k = 0, 1, \ldots. \tag{5.6}$$

and is called a *general secant method.*

The evaluation of $J(x^k, H_k)$ in the form (5.5) is rather costly. However, the next iterate x^{k+1} in (5.6) can be computed without explicit knowledge of the matrix $A_k = J(x^k, H_k)$. For this observe that, for $\{Fx^{k,j}\}$ in general position, the matrix of the $(n+1)$-dimensional system

$$\begin{pmatrix} 1 & 1 & \cdots & 1 \\ Fx^k & Fx^{k,1} & \cdots & Fx^{k,n} \end{pmatrix} \begin{pmatrix} z^0 \\ \vdots \\ z^n \end{pmatrix} = \begin{pmatrix} 1 \\ 0 \\ \vdots \\ 0 \end{pmatrix} \tag{5.7}$$

is nonsingular. Hence the unique solution satisfies

$$0 = \sum_{j=0}^{n} z_j F x^{k,j} = a^k + A_k \sum_{j=0}^{n} z_j x^{k,j}, \quad \sum_{j=0}^{n} z_j = 1,$$

which implies that

$$x^{k+1} = \sum_{j=0}^{n} z_j x^{k,j}.$$

Accordingly, a step of a secant method should, in general, be computed by the following algorithm.

$\mathcal{G}$: **input:** $\{k, x^k, M_k\}$
construct $x^{k,1}, \ldots, x^{k,n}$;
evaluate Fx^k, $Fx^{k,1}, \ldots, Fx^{k,n}$;
(5.8) solve (5.7) for $(z^0, \ldots, z^n)^\top$;
if solver failed **then return:** fail;
$x^{k+1} := z_0 x^k + \sum_{j=1}^{n} z_j x^{k,j}$;
return: $\{x^{k+1}, M_{k+1}\}$

The complexity of these methods depends strongly on the selection of the interpolating points, or, equivalently, on the choice of the matrices H_k. Note that the computation of the n^2 elements of $J(x^k, H_k)$ requires $n+1$ evaluations of F. It should also be noted that (5.7) has a unique solution if the $\{Fx^{k,j}\}$ are in general position, even if this is not true for the $\{x^{k,j}\}$. In such a case, x^{k+1} remains in the lower-dimensional affine subspace spanned by these points, and, if this persists, the iterates are unlikely ever to reach a solution.

Clearly, the general secant methods include the earlier introduced discretized Newton methods. For example, with the simple choice

$$H = \text{diag}\,(h_1, \ldots, h_n), \qquad h_1, \ldots, h_n \neq 0, \tag{5.9}$$

the mapping J represents the discretization (4.20) with $h_{ij} = h_j$, $1 \leq i, j \leq n$. Similarly, for

$$H = \begin{pmatrix} h_1 & h_1 & \ldots & h_1 \\ 0 & h_2 & \ldots & h_2 \\ \vdots & \vdots & \ddots & \vdots \\ 0 & 0 & \ldots & h_n \end{pmatrix} \tag{5.10}$$

it is readily seen that we obtain the case $\beta = 1$, $h_{ij} = h_j$, $1 \leq i, j \leq n$, of (4.21).

A natural choice of the parameters h_i^k in (5.9) and (5.10) is evidently $h_i^k = x_i^{k-1} - x_i^k$, $i = 1, \ldots, n$. The resulting methods then depend only on x^k and x^{k-1} and hence are sequential two-point processes. A general sequential $(p+1)$-point secant method may be generated by choosing, for example,

$$x^{k,j} = x^k + \sum_{i=1}^{p} P_{ijk}(x^{k-i} - x^k), \quad j = 1, \ldots, n, \quad P_{ijk} \in L(\mathrm{R}^n).$$

With $p = 1$ and $P_{ijk} = (0, \ldots, e^j, \ldots, 0)$ or $P_{ijk} = (e^1, \ldots, e^j, 0, \ldots, 0)$, this covers again the above cases (5.9) and (5.10), respectively. On the other hand, with $p = n$ and the identity matrix in place of each P_{ijk} we obtain the sequential $(n+1)$-point secant method specified by

$$H_k = (x^{k-1} - x^k, \ldots, x^{k-n} - x^k), \tag{5.11}$$

which offers considerable computational savings. In fact, now the step algorithm (5.8) requires, besides Fx^k, the vectors $Fx^{k-1}, \ldots, Fx^{k-n}$, which are already available from the prior steps. Thus, except at the start when $Fx^0, Fx^1, \ldots, Fx^n$ must all be calculated, only one new function evaluation is needed per step. Another computational saving is possible in the solution of the linear system (5.7). In fact, the corresponding matrices at the $(k-1)$th and kth iteration steps are

$$\begin{pmatrix} 1 & 1 & \ldots & 1 \\ Fx^{k-1} & Fx^{k-2} & \ldots & Fx^{k-n-1} \end{pmatrix}, \begin{pmatrix} 1 & 1 & \ldots & 1 \\ Fx^{k} & Fx^{k-1} & \ldots & Fx^{k-n} \end{pmatrix}$$

and hence, except for a column permutation, they differ only in one column. In other words, the difference between these two matrices is a matrix of rank one.

This can be used to compute the solution at step k by an update procedure along the line of the Sherman–Morrison–Woodbury formula without the need for a full decomposition of the full matrix. We shall return to this point in section 5.3.

A nonsequential $(n+1)$-point method can be generated by specifying the interpolating points, for instance, as those n among the previous iterates $x^0, \ldots, x^{k-1}$ for which $\|Fx^j\|$ is smallest. Another selection principle uses, in the definition of H_k, the function values Fx^j rather than the differences between previous iterates. This constitutes an n-dimensional extension of the classical one-dimensional Steffensen method. For example, with $h_i^k = f_i(x^k)$ in (5.9) and (5.10) we obtain simple one-point Steffensen methods, while

$$x^{k,j} = x^k + \sum_{i=1}^{p} P_{ijk} F x^{k-i+1}, \quad j = 1, \ldots, n, \quad P_{ijk} \in L(\mathrm{R}^n)$$

gives a sequential p-point Steffensen method.

Many other choices of interpolating points have been reported in the literature, including some in which the $x^{k,j}$ depend nonlinearly on previous iterates (see Robinson [216]). Note that in all cases we should guarantee that the $\{x^{k,j}\}$ and the $\{Fx^{k,j}\}$ are in general position. Unfortunately, except in special situations, this is rarely possible a priori.

The basic idea of the secant methods dates back to Gauss for the two-dimensional case (see Ostrowski [197]). An analysis of these methods in R^n was first given by Bittner [29]. Independently, Wolfe [281] suggested the sequential $(n+1)$-point method (5.11) and the computational approach incorporated in our step algorithm (5.8). A special form of a one-point Steffensen method was considered by Ludwig [158]. For some additional references see [OR].

5.2 Consistent Approximations

For the discretized Newton methods (4.19) and thus the more general secant methods (5.6) the attraction theory of section 3.3 is no longer applicable. In recent years only the basic discretized Newton methods appear to have found continued use, while interest in the more general secant methods has waned considerably. Accordingly, we present here only a simplified form of the general convergence theory developed in [OR].

Throughout this section it will be assumed that $F : E \subset \mathrm{R}^n \mapsto \mathrm{R}^n$ is a C^1 mapping on an open set E which satisfies, with constant $\gamma > 0$,

$$\|DF(y) - DF(x)\| \leq \gamma \|y - x\|, \quad \forall\, x, y \in E. \tag{5.12}$$

Clearly, for the development of a convergence theory of the secant methods we need to know how $J(x,h)$ approximates $DF(x)$. The theory of [OR] was based on a general consistency condition for this approximation, of which we use here only the following special case.

DEFINITION 5.1. *A mapping*

$$J : \; E_J \times E_h \mapsto L(\mathrm{R}^n), \quad E_J \subset E, \; E_h \subset \mathrm{R}^m \tag{5.13}$$

is a strongly consistent approximation *of DF if*

$$\|DF(x) - J(x,h)\| \le c\|h\|, \ \forall\ x \in E_J,\ h \in E_h(r) \equiv \{h \in E_h : \|h\| \le r\}$$

with some fixed positive constants c and r.

It is standard to show that reasonable discretizations of the Jacobian represent strongly consistent approximations. For example, the following result holds for (4.21).

THEOREM 5.2. *For any compact set $E_J \subset E$ there exists $r > 0$ such that with $E_h(r) = \{h \in \mathrm{R}^{n^2} : 0 < |h_{ij}| \le r,\ i,j = 1,\dots,n\}$ and $m = n^2$ the mapping (5.13) defined by (4.21) is a strongly consistent approximation of DF.*

Proof. By the compactness of E_J there exists $\delta > 0$ such that the compact set $C = \{x : \|y - x\|_1 \le \delta$ for some $y \in E_J\}$ is contained in E. With $r = \delta/n$ we have for any $h \in E_h(r)$, because $\beta \in [0,1]$,

$$\|\Delta_{i,j}(h) + h_{i,j}e^j\|_1 \le nr \le \delta, \quad \Delta_{i,j}(h) = \beta \sum_{k+1}^{j-1} h_{i,k}e^k,$$

whence $x + \Delta_{i,j}(h) + h_{i,j}e^j \in C$ for $x \in E_J$. By norm-equivalence on R^n, (5.12) implies that for some $\gamma_1 > 0$

$$\left|\frac{\partial}{\partial x_j} f_i(y) - \frac{\partial}{\partial x_j} f_i(x)\right| \le \gamma_1 \|y - x\|_1, \quad x, y \in E.$$

Thus, for $x \in E_J$, $h \in E_h(r)$, and with the abbreviation $y = x + \Delta_{i,j}(h)$ the integral mean-value theorem gives

$$\begin{aligned}
&\left|\frac{1}{h_{i,j}}(f_i(y + h_{i,j}e^j) - f_i(y)) - \frac{\partial}{\partial x_j} f_i(x)\right| \\
&\quad \le \int_0^1 \left|\frac{\partial}{\partial x_j} f_i(y + th_{ij}e^j) - \frac{\partial}{\partial x_j} f_i(y)\right| dt + \left|\frac{\partial}{\partial x_j} f_i(y) - \frac{\partial}{\partial x_j} f_i(x)\right| \\
&\quad \le \gamma_1 \left[|h_{i,j}| \int_0^1 t dt + |\Delta_{i,j}(h)| \right] \le \gamma_1 \sum_{k=1}^{n} |h_{i,k}|.
\end{aligned}$$

Hence

$$\|J(x,h) - DF(x)\|_1 \le \gamma_1 \sum_{i,k=1}^{n} |h_{i,k}| = \gamma_1 \|h\|_1. \ \square$$

This result together with the following local convergence theorem can be directly applied to discretized Newton methods.

THEOREM 5.3. *Let (5.13) be a strongly consistent approximation of DF on some open neighborhood $\mathcal{U} \subset E$ of a simple zero $x^* \in E$ of F. Then there exists a ball $\mathcal{B} = \bar{B}(x^*, \delta) \subset \mathcal{U}$, $\delta > 0$, and some $r > 0$ such that for $x^0 \in \mathcal{B}$ and any positive sequence $\{h^k\} \subset E_h(r) = E_h \cap B(0,r)$ the iterates*

$$x^{k+1} = x^k - J(x^k, h^k)^{-1} F x^k, \quad k = 0, 1, \dots \tag{5.14}$$

are well defined, remain in $\mathcal{B}$, and converge to x^*. If $\lim_{k\to\infty} h^k = 0$, then $R_1\{x^k\} = Q_1\{x^k\} = 0$.

Proof. Set $\beta = \|DF(x^*)^{-1}\|$ and choose $\delta > 0$ and $r > 0$ such that $\mathcal{B} = \bar{B}(x^*, \delta) \subset \mathcal{U}$ and $\epsilon = \frac{3}{2}\gamma\delta + cr < \frac{1}{2\beta}$. Let $x \in \mathcal{B}$ and $h \in E_h(r)$. Then

$$\begin{aligned} \|DF(x^*) - J(x,h)\| &\leq \|DF(x^*) - DF(x)\| + \|DF(x) - J(x,h)\| \\ &\leq \gamma\|x - x^*\| + c\|h\| = \gamma\delta + cr < \epsilon < \frac{1}{2\beta} \end{aligned}$$

implies the invertibility of $J(x,h)$ and $\|J(x,h)^{-1}\| \leq \eta$ with $\eta = \beta/(1-\beta\epsilon)$. Hence the mapping $G : \mathcal{B} \times E_h(r) \mapsto \mathrm{R}^n$, $G(x,h) = x - J(x,h)^{-1}Fx$ is well defined and

$$\begin{aligned} \|G(x,h) - x^*\| &= \|J(x,h)^{-1}[J(x,h)(x-x^*) - Fx]\| \\ &\leq \eta\Big[\|J(x,h) - DF(x)\| + \|DF(x) - DF(x^*)\|\Big]\|x - x^*\| \\ &\qquad + \eta\|Fx - Fx^* - DF(x^*)(x-x^*)\| \\ &\leq \eta\left[c\|h\|\,\|x - x^*\| + \frac{3}{2}\gamma\|x - x^*\|^2\right] \leq \alpha\|x - x^*\| \end{aligned} \tag{5.15}$$

with $\alpha \leq \eta[3\gamma\delta/2 + cr] = \beta\epsilon < 1$. This shows that, for $x^0 \in \mathcal{B}$ and $\{h^k\} \subset E_h(r)$, the iterates $x^{k+1} = G(x^k, h^k)$ remain in $\mathcal{B}$ and satisfy $\|x^{k+1} - x^*\| \leq \alpha\|x^k - x^*\|$ for all $k \geq 0$, which implies their convergence to x^*. Moreover if $h^k \to 0$, then

$$R_1\{x^k\} \leq Q_1\{x^k\} \leq \eta \limsup_{k\to\infty} \left[c\|h^k\| + \frac{3}{2}\gamma\|x^k - x^*\|\right] = 0$$

as claimed. □

With appropriate choices of the h^k the rate of convergence can be improved. For this note that, under the conditions of Theorem 5.3, we have $Fx^k = 0$ for sufficiently large k only if $x^k = x^*$, since x^* is the only zero of F in some ball around that point. Moreover, the invertibility of $J(x^k, h^k)$ implies that $x^{k+1} = x^k$ if and only if $Fx^k = 0$. Thus, either the process (5.14) stops at x^* after finitely many steps or we have $Fx^k \neq 0$ and $x^{k+1} \neq x^k$ for all $k \geq 0$.

COROLLARY 5.4. *Under the conditions of Theorem* 5.3 *suppose that the steps $h^k \in E_h(r)$ are chosen such that, for some $k_0 \geq 0$,*

$$\|h^k\| \leq \beta_1\|Fx^k\|, \quad \forall\, k \geq k_0, \tag{5.16}$$

as long as the process has not stopped at x^*. Then $O_R\{x^k\} \geq O_Q\{x^k\} \geq 2$. Analogously, if

$$\|h^k\| \leq \beta_2\|x^k - x^{k-1}\|, \quad \forall\, k \geq k_0, \tag{5.17}$$

then $O_R\{x^k\} \geq 1/2(1+\sqrt{5})$.

The proof for the choice (5.16) follows directly from the estimate (5.15), since then

$$\|x^{k+1} - x^*\| \leq \eta\left[c\beta_1 + \frac{3}{2}\gamma\right]\,\|x^k - x^*\|^2.$$

For the proof in the case of (5.17) we refer to [OR].

The conditions (5.16), (5.17) appear to have been used first by Samanskii [221]. Voigt [270] exhibits various special conditions under which the orders of convergence are actually equal to the cited values. Theorems 5.3 and 5.4 are used in [OR] to prove convergence of several of the two-point secant and one-point Steffensen processes discussed in section 6.1. For discretized Newton methods, Theorem 5.3 suggests the use of, say, (4.21) with $h_{1,j} = \cdots = h_{n,j} = h_j$ given by $h_j = \min\,(h_{\max}, \max\,(f_j(x^k), h_{\min}))$ where the bounds $h_{\max} \geq h_{\min} > 0$ are chosen suitably, depending on the floating-point arithmetic. Alternatively, we may use $h_j = \min\,(h_{\max}, \max\,(x_j^k - x_j^{k-1}, h_{\min}))$.

Various other forms of convergence theorems have been given in the literature. Schmidt [229] and several other authors assume that J is a general divided-difference operator in the sense that $J(x,h)h = F(x+h) - Fx$. Under certain additional norm conditions, this permits the proof of semilocal convergence results for several of these methods.

These results also raise the question of when the mapping J of (5.5) used in the general secant method (5.6) is a strongly consistent approximation of DF. Clearly, this requires some restriction on the class of matrices H_k. A set of matrices $Q \subset L(\mathrm{R}^n)$ is *uniformly nonsingular* if it is a subset of

$$K(\sigma) = \left\{ H = (h^1, \ldots, h^n) \in L(\mathrm{R}^n) \ : \ h^j \neq 0,\ \forall\, j, \left| \det\left(\frac{h^1}{\|h^1\|}, \ \ldots, \ \frac{h^n}{\|h^n\|} \right) \right| \geq \sigma \right\}$$

for some constant $\sigma > 0$. In [OR] it is shown that a sufficient condition for this is $\|H\|\ \|H^{-1}\| \leq \beta$ for all $H \in Q$ and that the following result holds.

THEOREM 5.5. *Let $Q \subset L(\mathrm{R}^n)$ be a uniformly nonsingular family with the zero matrix as limit point. Then for any compact set $E_C \subset E$ there exists $\delta > 0$ such that the mapping* (5.13) *specified by* (5.5) *is well defined for $E_J = E_C$, $E_h = \{H \in Q \ : \ \|H\| \leq \delta\}$, and is a strongly consistent approximation of DF.*

With this we may apply Theorems 5.3 and 5.4 to the general secant methods. As a special example we cite a result about the sequential $(n+1)$-point secant method defined by the matrices (5.11).

THEOREM 5.6. *Let $x^* \in E$ be a simple zero of F. Suppose that the sequence $\{x^k\} \subset E$ generated by* (5.6), (5.11) *is well defined, $\|x^0 - x^*\|$ is sufficiently small, and that $H_k \in \{H \in K(\sigma) \ : \ \|H\| \leq \delta\}$ for some $\sigma > 0$, $\delta > 0$. Then $\lim_{k\to\infty} x^k = x^*$ and $O_R\{x^k\} \geq \tau$, where τ is the positive root of $t^{n+1} - t^n - 1 = 0$.*

This is a very weak result in that it assumes the $\{x^k\}$ to be well defined and the $\{H_k\}$ to remain in some $K(\sigma)$. Only the rate-of-convergence statement may be of some interest, although it should be noted that τ depends on the dimension n. In fact, τ tends to one when n grows larger, and, for instance, for $n = 100$ we have $\tau \doteq 1.03$. The condition $H_k \in K(\sigma)$ is very severe and under simple assumptions about F, it can be shown (see [OR]) that it is possible to find, for any $\delta > 0$ and $\epsilon > 0$, a nonsingular H with $\|H\| \leq \delta$ and

$\|DF(x) - J(x, H)\| \geq \epsilon$. This means that the sequential $(n + 1)$-point secant method has no guaranteed local convergence.

5.3 Update Methods

As noted, the $(n + 1)$-point sequential secant method has computationally advantageous properties but is, in general, not locally convergent. After the initial phase, each step of the method requires only one evaluation of the given mapping, and the matrices of the linear systems at succeeding steps differ by a rank-one matrix. Moreover, the order of convergence exceeds one—provided there is convergence. It is natural to ask whether there are linearization methods which have similar properties but show a better convergence behavior. In particular, we are interested in the computationally advantageous property that succeeding iteration matrices differ only by a matrix of low rank.

For the solution of the equation $Fx = 0$, the $(n + 1)$-point sequential secant process is a linearization method (4.2) with

$$A_k = (Fx^k - Fx^{k-1}, \cdots, Fx^{k-n+1} - Fx^{k-n})(x^k - x^{k-1}, \ldots, x^{k-n+1} - x^{k-n})^{-1}.$$

For the differences $\Delta A_k = A_{k+1} - A_k$ this implies that

$$\begin{aligned} &(5.18) \qquad & \Delta A_k(x^{j+1} - x^j) &= 0, \quad j = k-1, \ldots, k-n+1, \\ &(5.19) \qquad & A_{k+1}(x^{k+1} - x^k) &= Fx^{k+1} - Fx^k, \end{aligned}$$

whence, by subtraction of the defining equation $A_k(x^{k+1} - x^k) + Fx^k = 0$ of the iterates, it follows that

$$\Delta A_k(x^{k+1} - x^k) = Fx^{k+1}. \qquad (5.20)$$

Since $x^{k+1}, \ldots, x^{k-n+1}$ must be in general position, (5.18) and (5.20) prove once again that ΔA_k is a rank-one matrix, and, more specifically, that $\Delta A_k = b^k(c^k)^\top$, where b^k is a multiple of Fx^{k+1} and c^k is orthogonal to $x^k - x^{k-1}, \ldots, x^{k-n+2} - x^{k-n+1}$. As shown by Barnes [19] this characterizes the method.

Thus, if other linearization methods are to be considered with the computationally desirable property that ΔA_k is a rank-one matrix, then the vectors b^k or c^k in $\Delta A_k = b^k(c^k)^\top$ should be selected differently. In some sense, the local convergence difficulties of the $(n+1)$-point sequential secant method are attributable to the requirement that c^k is orthogonal to $x^k - x^{k-1}, \ldots, x^{k-n+2} - x^{k-n+1}$. This suggests that we drop the conditions (5.18). On the other hand, in the one-dimensional case, (5.19) defines A_{k+1} as the standard forward difference quotient. Hence, if we want to keep some relationship with the secant methods, it is desirable to retain this generalized *divided-difference condition* (5.19). Then also (5.20) holds and thus b^k is a multiple of Fx^{k+1}.

In a slight extension of these ideas, and with a historically prompted change in notation, we now consider processes of the form

$$(5.21) \qquad \begin{aligned} &\text{(a)} \quad x^{k+1} = x^k - \alpha_k B_k^{-1} Fx^k, \\ &\text{(b)} \quad B_{k+1}(x^{k+1} - x^k) = Fx^{k+1} - Fx^k, \quad k = 0, 1, \ldots, \\ &\text{(c)} \quad B_{k+1} = B_k + \Delta B_k, \quad \operatorname{rank} \Delta B_k = m \geq 1. \end{aligned}$$

The fixed rank m is, in practice, at most two, and the relaxation factors α_k were introduced in anticipation of the application of the methods to minimization problems. We call (5.21) a *direct update method of rank* m. Many authors tend to consider the B_k as approximations to the derivatives $DF(x^k)$ and accordingly the term *quasi-Newton method* was suggested by Broyden [40] and is frequently used. This name is not particularly descriptive, since the same consideration might be applied to all linearization methods.

The processes (5.21) are well defined only if the inverses $B_k^{-1} = H_k$ exist for all k. Then the Sherman–Morrison–Woodbury formula applies and provides an explicit expression for $\Delta H_k = H_{k+1} - H_k$ and also shows that rank $\Delta H_k = m$. With this, the methods may be rewritten in the form

$$(5.22)\qquad \begin{aligned} &\text{(a)} \quad x^{k+1} = x^k - \alpha_k H_k F x^k, \\ &\text{(b)} \quad H_{k+1}(Fx^{k+1} - Fx^k) = x^{k+1} - x^k, \quad k = 0, 1, \ldots, \\ &\text{(c)} \quad H_{k+1} = H_k + \Delta H_k, \quad \operatorname{rank} \Delta H_k = m \geq 1. \end{aligned}$$

Whether they were derived from (5.21) in this way or not, the processes (5.22) are called *inverse update methods of rank* m. If $H_k^{-1} = B_k$ always exists, then (5.21) and (5.22) are said to be inverses of each other. Note that when (5.22) is considered by itself, then the H_k need not be invertible for the method to be well defined. Yet in most cases this destroys the hope of a general convergence result (see, e.g., Broyden [43]).

For the discussion of specific cases of the update formulas (5.21), (5.22) it is useful to delete all indices and to use the notation

$$(5.23)\qquad \left\{ \begin{array}{l} \bar{x} = x^{k+1},\ x = x^k,\ s = \bar{x} - x,\ y = F\bar{x} - Fx, \\ \bar{B} = B_{k+1},\ \bar{H} = H_{k+1},\ B = B_k,\ H = H_k, \\ \Delta B = \bar{B} - B,\ \Delta H = \bar{H} - H, \\ \Delta B s = u \equiv y - Bs,\ \Delta H y = v \equiv s - Hy, \end{array} \right.$$

where $\Delta B s = u$ and $\Delta H y = v$ incorporate the divided-difference condition (5.21)(b). In the rank-one case $\Delta B = bc^\top$, the condition $\Delta B s = u$ requires that $u = (c^\top s)b$. Clearly, u should not be restricted to zero, and thus we define the admissible direct update formulas of rank one by

$$(5.24)\qquad \Delta B = \frac{uc^\top}{c^\top s} \quad \text{if } c^\top s \neq 0, \quad \text{else } \Delta B = 0.$$

Correspondingly, the admissible inverse update formulas of rank one are given by

$$(5.25)\qquad \Delta H = \frac{vd^\top}{d^\top y} \quad \text{if } d^\top y \neq 0, \quad \text{else } \Delta H = 0.$$

If, in (5.24), $B^{-1} = H$ exists and we assume $c^\top s \neq 0$, then the Sherman–Morrison–Woodbury formula ensures that $\bar{B}$ is nonsingular exactly if $c^\top H y \neq 0$, in which case $\bar{H} = \bar{B}^{-1}$ satisfies (5.25) with $d = H^\top c$.

Various choices of c or d have been considered in the literature, including $c = y$ (see Pearson [200]), $c = u$ or, equivalently, $d = v$ (see Broyden [40]), $d = s$

(see Pearson [200]), and $d = y$ (see Broyden [39]). Experience has shown that the most successful choice is $c = s$ which leads to the *Broyden update* formula

$$\bar{B} = B + \frac{1}{s^\top s}(y - Bs)s^\top \tag{5.26}$$

given in Broyden [39]. It is characterized by the following *least-change* property.

THEOREM 5.7. *For any $y, s \in \mathrm{R}^n$ and $B \in L(\mathrm{R}^n)$, (5.26) is the unique solution of*

$$\min(\|\bar{B} - B\|_F \ : \ \bar{B} \in L(\mathrm{R}^n),\ \bar{B}s = y). \tag{5.27}$$

Proof. With $u = y - Bs$ and $\Delta B = \bar{B} - B$, (5.27) is equivalent to the minimization of $\|\Delta B\|_F$ over all $\Delta B \in L(\mathrm{R}^n)$ such that $\Delta B s = u$. Set $V = us^\top / s^\top s$, then any such ΔB decomposes as $\Delta B = V + C$ where, because of $Vs = u$, it follows that $Cs = 0$. For the Frobenius inner product (1.3) we obtain $\langle V, C\rangle_F = (1/\|s\|_2{}^2)\text{trace } (Cs)u^\top = 0$, whence

$$\|\Delta B\|_F{}^2 = \langle V + C, V + C\rangle_F = \langle V, V\rangle_F + \langle C, C\rangle_F = \frac{\|u\|_2{}^2}{\|s\|_2{}^2} + \|C\|_F{}^2.$$

This proves the result. □

For a detailed presentation of update formulas with the least-change property we refer to Dennis and Schnabel [65], and Dennis and Walker [68], [69].

The variety of possible methods increases considerably in the rank-two case. For unconstrained minimization problems, update formulas which preserve the symmetry of the matrices are desirable, and we restrict ourselves to this case. Then the direct updates are of the form

$$\Delta B = \begin{pmatrix} b & c \end{pmatrix} \Sigma \begin{pmatrix} b^\top \\ c^\top \end{pmatrix}, \quad \Sigma = \begin{pmatrix} \sigma_1 & \sigma_2 \\ \sigma_2 & \sigma_3 \end{pmatrix}, \quad b, c \in \mathrm{R}^n. \tag{5.28}$$

If the matrix Σ is definite then $s^\top \Delta B s = u^\top s$ requires that $u^\top s$ must always have constant sign. Already, simple examples show that this would lead to convergence problems. In fact, say, for F equal to the identity map on R^2, we have $u^\top s = s^\top(I_2 - B)s$, and it is easily seen that there exist symmetric matrices B close to I_2 and vectors x close to zero such that $u^\top s$ is either positive, negative, or even zero. The symmetric matrix Σ is indefinite exactly if its determinant $\delta_0 = \det \Sigma$ is nonpositive. It is reasonable to require that Σ is nonsingular and hence $\delta_0 < 0$.

Since the vectors b, c are essentially free, some suitable basis in R^2 may be chosen in which Σ assumes a simpler form. In particular, for $\delta_0 < 0$ we may transform Σ such that either σ_1 or σ_3 is zero. In fact, if, say, $\sigma_3 \neq 0$ then a simple calculation shows that

$$\Sigma = \begin{pmatrix} 1 & \mu \\ 0 & 1 \end{pmatrix} \begin{pmatrix} 0 & \delta_1 \\ \delta_1 & \sigma_3 \end{pmatrix} \begin{pmatrix} 1 & 0 \\ \mu & 1 \end{pmatrix}, \ \mu = \frac{\sigma_2 - \delta_1}{\sigma_3}, \ \delta_1 = \sqrt{-\delta_0}.$$

Thus, when $\delta_0 < 0$, there is no loss of generality to assume that $\sigma_1 = 0$ in (5.28). The condition $\Delta B s = u$ requires u to be in the subspace spanned by b and c

and hence it is no restriction to set $b = u$. Then, for any $c \in \mathrm{R}^n$ such that $c^\top s \neq 0$ it follows that $\sigma_2 = 1/c^\top s$ and $\sigma_3 = -u^\top s/(c^\top s)^2$. In other words, all symmetric direct update formulas with nonpositive determinants can be written in the form

$$\Delta B = \frac{uc^\top + cu^\top}{c^\top s} - \frac{u^\top s}{(c^\top s)^2}\, cc^\top, \quad \text{if } c^\top s \neq 0, \text{ else } \Delta B = 0. \tag{5.29}$$

Powell [205] (see also Dennis [64]) applied an iterative argument to (5.24) to produce these symmetric rank-two formulas. Accordingly, we follow Powell's terminology and call (5.29) the *symmetrization* of (5.24).

For $c = s$ (5.29) becomes the *Powell-symmetric–Broyden* (PSB) update formula

$$\bar{B} = B + \frac{us^\top + su^\top}{s^\top s} - \frac{s^\top u\; ss^\top}{(s^\top s)^2} \tag{5.30}$$

of Powell [206] while, for $c = y$ we obtain the *Davidon–Fletcher–Powell* (DFP) update formula

$$\bar{B} = B + \frac{uy^\top + yu^\top}{y^\top s} - \frac{s^\top u\; yy^\top}{(y^\top s)^2} \tag{5.31}$$

given in Davidon [61] and Fletcher and Powell [97]. Analogous to the Broyden update (5.26), the PSB formula (5.30) has the following least-change property.

THEOREM 5.8. *For any $y, s \in \mathrm{R}^n$ and $B \in L_S(\mathrm{R}^n)$, the matrix $\bar{B}$ of (5.30) is the unique solution of*

$$\min\, (\|\bar{B} - B\|_F\ ;\ \bar{B} \in L_S(\mathrm{R}^n),\ \bar{B}s = y). \tag{5.32}$$

Proof. The proof is the same as that of Theorem 5.7. With $u = y - Bs$ and $\Delta B - \bar{B} - B$, (5.32) requires minimizing $\|\Delta B\|_F$ over all $\Delta B \in L_S(\mathrm{R}^n)$ such that $\Delta Bs = u$. Set

$$V = \frac{us^\top + su^\top}{s^\top s} - \frac{s^\top u}{(s^\top s)^2}\, ss^\top, \quad \Delta B = V + C,\ C \in L_S(\mathrm{R}^n).$$

Then $Vs = u$ and hence $Cs = 0$. As in the proof of Theorem 5.7, it follows that $\langle V, C\rangle_F = 0$, whence

$$\|\Delta B\|_F^2 = \langle V, V\rangle_F + \langle C, C\rangle_F = \frac{\|u\|_2^2}{\|s\|_2^2}\left[2 - \frac{(s^\top u)^2}{\|u\|_2^2\|s\|_2^2}\right] + \|C\|_F^2. \ \square$$

For (5.31) the same argument does not lead to the orthogonality condition and to a corresponding least-change result. In such cases a scaled inner product can be used, as, for instance,

$$A, B \in L(\mathrm{R}^n) \ \mapsto\ \text{trace}\ ((W_1 A W_2)^\top (W_1 B W_2))$$

with invertible $W_1, W_2 \in L(\mathrm{R}^n)$. For (5.31) the following result holds.

THEOREM 5.9. *For any* $y, s \in \mathrm{R}^n$, *and positive-definite* $B \in L_S(\mathrm{R}^n)$, *the matrix* $\bar{B}$ *of* (5.31) *is the unique solution of*

$$(5.33) \qquad \min \left(\|W(\bar{B} - B)W\|_F \ ; \ \bar{B} \in L_S(\mathrm{R}^n) \ \bar{B}s = y \right)$$

with $W = B^{-1/2}$.

The proof is analogous to that of Theorems 5.7 and 5.8. Under the conditions of the theorem, $\bar{B}$ will again be positive definite provided that $s^\top y > 0$ (see, e.g., section 8.2.3 of [OR]).

The development of inverse rank-two formulas can be handled analogously. More specifically, it follows that all inverse symmetric rank-two formulas with negative determinant can be written in the form

$$(5.34) \qquad \Delta H = \frac{vd^\top + dv^\top}{d^\top y} - \frac{v^\top y}{d^\top y} \frac{dd^\top}{d^\top y}, \quad \text{if } d^\top y \neq 0, \text{ else } \Delta H = 0,$$

with some free parameter vector $d \in \mathrm{R}^n$. This is called the symmetrization of (5.25).

For $d = Hc$; that is, when (5.25) is the inverse of (5.24), one might expect that (5.34) and (5.29) are also inverses of each other. In general, this is not the case. In fact, if H and $\bar{H}$ are symmetric and invertible then an application of the Sherman–Morrison–Woodbury formula shows, after a simple calculation, that $\Delta B = \bar{H}^{-1} - H^{-1}$ satisfies

$$(5.35) \qquad \Delta B = \frac{(c^\top d)uu^\top + c^\top s(uc^\top + cu^\top) - (u^\top s)cc^\top}{(c^\top s)^2 + (u^\top s)(c^\top d)}, \quad c = Bd.$$

This is not of the form (5.29). The formulas (5.29) with some $c \in \mathrm{R}^n$ and (5.34) with $d = Hc$ are often called dual or complementary to each other.

For the choices $d = y$ and $d = Hy$, (5.34) reduces to two formulas of Greenstadt [117]. With $d = s$, (and hence c = Bs), the inverse (5.35) of (5.34) is a formula independently suggested by Broyden [41], Fletcher [95], Goldfarb [111], and Shanno [244], the so-called *BFGS formula*

$$(5.36) \qquad \bar{B} = B + \frac{yy^\top}{y^\top s} - \frac{Bss^\top B}{s^\top Bs}.$$

This is widely considered the most effective update formula for minimization problems.

It should be noted that in our discussion of the rank-two formulas we did not use the linear independence of u and c (or v and d). The choice $c = u$ (or $d = v$) is therefore permitted in (5.29) (or (5.34)), but the result is simply the rank-one formula (5.24) (or (5.25)) with the same parameter vectors.

The form of the update schemes discussed so far is strongly influenced by the validity of the divided-difference equations (5.21)(b), (5.22)(b). There was no inherent reason why these equations should be required except to retain some connection with the secant methods. But that relationship would not really be broken if we were to use instead of, say, (5.22)(b), the condition

$$(5.37) \qquad H_{k+1}(Fx^{k+1} - Fx^k) = \rho_k(x^{k+1} - x^k), \quad k = 0, 1, \ldots$$

with real parameters ρ_k. This was done by Huang [135] who developed a class of inverse update formulas which intersects the update class (5.34), but is neither a subset nor a superset of it.

From a computational viewpoint inverse update formulas appear to have the advantage of requiring at each step only a matrix-vector multiplication rather than the solution of a linear system. However, the complexity of the direct methods can be improved by updating the Cholesky factors of the matrices. Briefly, suppose that both $A_0 \in L_S(\mathrm{R}^n)$ and $A_1 = A_0 + \sigma w w^\top \in L_S(\mathrm{R}^n)$ are positive definite. If $A_0 = L_0 L_0^\top$ (L_0 lower triangular) is the Cholesky decomposition of A_0, then with the solution z of $L_0 z = w$ we obtain $A_1 = L_0(I + \sigma z z^\top)L_0^\top$. Hence, it suffices to compute the Cholesky decomposition $I + \sigma z z^\top = \hat{L}\hat{L}^\top$ to obtain the corresponding decomposition $A_1 = L_1 L_1^\top$ with $L_1 = L_0 \hat{L}$.

For symmetric rank-two formulas the process has to be applied twice. But, as shown by Goldfarb [112], for the BFGS update formula (5.36) this can be simplified. In fact, suppose that $B \in L_S(\mathrm{R}^n)$ is positive definite and has the Cholesky decomposition $B = LL^\top$ and that $y^\top s > 0$. Then, it can be proved that the matrix $\bar{B}$ of (5.36) is again positive definite (see, e.g., Gill, Murray, and Wright [109, p.120]). Moreover, a short calculation shows that $\bar{B}$ satisfies

$$\bar{B} = KK^\top, \quad K = L + \frac{1}{y^\top s}(y - Lw)w^\top, \; w = \left[\frac{y^\top s}{s^\top B s}\right]^{\frac{1}{2}} L^\top s.$$

Hence, we need to compute only the QR-factorization $K^\top = Q\bar{L}^\top$ of K (without storing Q) to obtain the Cholesky decomposition $\bar{B} = \bar{L}\bar{L}^\top$ of $\bar{B}$. The required factorization of $K^\top$ can be accomplished with $\mathcal{O}(n^2)$ operations (see Golub and Van Loan [114, 12.6]). This has become the preferred way of implementing the BFGS update formula when all matrices can be stored in full.

The use of the Cholesky factorization for the implementation of update methods was introduced in Gill and Murray [107] and Goldfarb [112] (see also Fletcher and Powell [98], Gill, Golub, Murray, and Saunders [106], Gill, Murray, and Saunders [108], and Brodlie [33]).

In the case of large problems with sparse Jacobians the inverse Jacobians are generally not sparse and the same is expected of the matrices arising in inverse update formulas. However, even for direct formulas the situation is not so simple. In fact, even if B_k is sparse, the increment ΔB_k generally does not share this property and hence the matrices tend to fill up quickly during the iteration. Schubert [235] and Broyden [42] proposed the following technique for overcoming this problem. Let T_F be the pattern matrix of F (see Definition 2.1). For the update step $B_{k+1} = B_k + \Delta B_k$, we replace in ΔB_k by zeros all elements that correspond to zero entries of T_F. Then each row of the resulting matrix $\tilde{\Delta} B_k$ is scaled so that the divided-difference condition (5.21)(b) once again holds. More specifically, $\Delta \tilde{B}_k$ is multiplied by a diagonal matrix D_k such that $\tilde{B}_{k+1} = B_k + D_k \tilde{\Delta} B_k$ satisfies $\tilde{B}_{k+1} s^k = y^k$. Now the process continues with $\tilde{B}_{k+1}$ in place of B_{k+1}.

Dennis and Schnabel [65], as well as Dennis and Walker [68], [69] analyzed this approach in the context of a convergence theory for least-change update methods. The Q-superlinear convergence of Broyden's method was established by Broyden, Dennis, and Moré [44] with a proof procedure that has become typical for least-change update methods.

THEOREM 5.10. *Let* $F : E \subset \mathbb{R}^n \mapsto \mathbb{R}^n$ *be* C^1 *on an open set* E *with Lipschitz-continuous first derivative. Then, for any simple zero* $x^* \in E$ *of* F, *there exist* $\delta, \epsilon > 0$ *such that for* $x^0 \in B(x^*, \epsilon)$ *and* $B_0 \in L(\mathbb{R}^n)$ *with* $\|B_0 - DF(x^0)\| < \delta$ *the iteration* (5.21) *with the Broyden update* (5.26) *is well defined and converges Q-superlinearly to* x^*.

For the proof and the general convergence theory of least-change update methods we refer to the cited articles.

CHAPTER 6

Combinations of Processes

This chapter addresses the analysis of combined iterative processes where at each step of a primary method a secondary iteration is applied. For example, when linearization methods are used for solving high-dimensional problems it may become necessary to work with a secondary linear iterative process to solve the resulting linear systems at each step. In the study of such combined processes we have to distinguish between controlling the secondary method by means of a priori specified conditions or by some adaptive strategies based on the performance of the computation. We shall follow established terminology and identify the various combined processes considered here by concatenating the names of the constituent primary and secondary methods.

6.1 The Use of Classical Linear Methods

At each step of a linearization method (4.2) we have to solve a linear system

$$A_k y = Fx^k. \tag{6.1}$$

In this section we consider the use of certain classical linear iterative methods, such as the SOR or ADI processes, to solve (6.1). The results discussed here are nonadaptive in the sense that the number of secondary steps is either a priori given or that the secondary process is run to convergence. The resulting combined methods are again linearization methods for which the convergence theorems of chapter 4 are applicable. But the nonadaptive control of the secondary process is restrictive and causes these results not to be very practical. Hence our presentation will be brief. Adaptive controls will be discussed beginning with section 6.3.

Many of the classical linear iterative processes, when applied to the solution of (6.1), are constructed from a splitting $A_k = B_k - C_k$ with nonsingular B_k by setting

$$y^{j+1} = H_k y^j + B_k^{-1} Fx^k, \quad j = 0, 1, \ldots, \quad H_k = B_k^{-1} C_k. \tag{6.2}$$

Since we expect $Fx^k \to 0$ as $k \to \infty$, it is natural to start with $y^0 = 0$. If (6.2) is stopped after m_k steps; that is, if $x^{k+1} = y_{m_k}$ is taken as the next primary iterate, then the combined process is readily seen to have the form

$$x^{k+1} = x^k - \left(I + H_k + \cdots + H_k^{m_k - 1}\right) B_k^{-1} Fx^k, \quad k = 0, 1, \ldots. \tag{6.3}$$

The secondary method (6.2) converges for any starting vector if and only if $\rho(H_k) < 1$. In that case, because $(I + H + \cdots + H^{m-1})(I - H) = I - H^m$, we see that the matrix

$$\hat{A}_k = B_k[I + H_k + \cdots + H_k^{m_k-1}]^{-1}$$

is nonsingular, and hence that (6.3) is again a linearization method.

As a typical example of (6.3) consider Newton's method as the primary process and let $DF(x) = D(x) - L(x) - U(x)$ be the usual decomposition of the matrix $DF(x)$ into its diagonal and triangular parts. If $D(x)$ is nonsingular, then the splitting $DF(x) = B(x) - C(x)$, with $B(x) = \omega^{-1}[D(x) - \omega L(x)]$, defines the SOR iteration for the linear system $DF(x)y = b$. Its iteration matrix has the form

$$H_\omega(x) = [D(x) - \omega L(x)]^{-1}[(1-\omega)D(x) + \omega U(x)], \tag{6.4}$$

and with it (6.3) becomes the Newton–SOR method

$$\begin{aligned} x^{k+1} = x^k - {} & \omega_k[I + H_{\omega_k}(x^k) + \ldots + H_{\omega_k}(x^k)^{m_k-1}] \\ & \Big[D(x^k) - \omega_k L(x^k)\Big]^{-1} Fx^k, \qquad k = 0, 1, \ldots. \end{aligned} \tag{6.5}$$

A related example is the (relaxed) one-step Newton–Jacobi method

$$x^{k+1} = x^k - \omega_k D(x^k)^{-1} Fx^k, \qquad k = 0, 1, \ldots. \tag{6.6}$$

Instead of a secondary linear iteration based on a splitting of the matrix A_k of (6.1) we may also apply an alternating direction (ADI) method. If, for instance, only one ADI step is used, then the resulting combined method has the form

$$x^{k+1} = x^k - 2\alpha[V_k + \alpha I]^{-1}[H_k + \alpha I]^{-1}Fx^k, \ k = 0, 1, \ldots,$$

where now $A_k = H_k + V_k$ and all inverses are assumed to exist. Formulation of the corresponding m-step case should be clear.

The application of the SOR iteration as a secondary process has been standard for many years. Various results were given by Greenspan and Yohe [118], Greenspan [115], Greenspan and Parter [116], Ortega and Rockoff [194], Moré [175], and [OR]. The use of the ADI method as the secondary iteration for discrete analogues $Ax = \Phi x$ of mildly nonlinear boundary-value problems was discussed by Douglas [81], [82], [83]. In that case, the primary method was of the Picard type $(A+\gamma I)x^{k+1} = \Phi x^k + \gamma x^k$. A similar case was studied by Gunn [120].

General local convergence results for these combined processes are only available for primary processes of Newton form (4.11). In that case, under the conditions of Theorem 4.3, the local convergence of the primary method is ensured if $\sigma = \rho(I - A(x^*)^{-1}DF(x^*)) < 1$. Suppose that as a secondary process we use the method defined by a splitting $A(x) = B(x) - C(x)$ for which the mapping $B : \mathcal{U} \mapsto L(\mathrm{R}^n)$ is continuous at x^* with nonsingular $B(x^*)$ and

$\rho(B(x^*)^{-1}C(x^*)) < 1$. Then there exists an open neighborhood $\mathcal{U}_0 \subset \mathcal{U}$ of x^* such that for any x in $\mathcal{U}_0$ and fixed $m \geq 1$ the matrices $H(x) = B(x)^{-1}C(x)$ and

$$\hat{A}(x) = B(x)[I + H(x) + \cdots + H(x)^{m-1}]^{-1}, \qquad m \geq 1 \tag{6.7}$$

are well defined (see section 10.3 of [OR]). Moreover, $\hat{A}$ is continuous at x^* with nonsingular $\hat{A}(x^*)$ and, therefore, Theorem 4.3 holds for $\hat{A}$. The local convergence of the combined process (6.3) with constant $m_k = m > 1$ for all $k \geq 0$ is then guaranteed if

$$\hat{\sigma} = \rho(I - \hat{A}(x^*)DF(x^*)) = [\rho(H(x^*))]^m < 1, \tag{6.8}$$

in which case the R_1-factor of the process equals $\hat{\sigma}$. Thus, for $m = 1$ the rate of convergence is exactly equal to that of the secondary linear process applied to

$$DF(x^*)y = b, \tag{6.9}$$

and the m-step process is m times as fast.

The condition $\rho(I - A(x^*)^{-1}DF(x^*)) < 1$ for the primary process is exactly the convergence criterion for the linear iteration defined by the splitting $DF(x^*) = A(x^*) - (A(x^*) - DF(x^*))$ for solving (6.9).

Correspondingly, if, say, $m = 1$, then the condition (6.8) for the combined process controls the convergence of the iteration for (6.9) given by $DF(x^*) = B(x^*) - (A(x^*) - DF(x^*) + C(x^*))$. Thus, if, for example, the theory of regular splittings applies (see, e.g., Varga [269, p. 87]), then the rate of convergence of the combined process is expected to be worse than that of the original method. This is particularly striking if the primary process is Newton's method itself; that is, if $\sigma = 0$; while by (6.8), for any constant $m \geq 1$ the combined process converges only R-linearly.

These results also show that the rates of convergence of these methods become worse when the dimension increases. In fact, as shown, e.g., by Varga [269], the (point) Jacobi process applied to an n-dimensional linear system has the R-factor $R_1 = \cos(\pi/n)$ which tends to one as $1/n \to 0$. Thus the same result will hold for a combined process with the Jacobi method as the secondary iteration. A similar result holds for the SOR method as the secondary iteration.

When the assumption of constant $m_k = m$ for all $k \geq 0$ is dropped the basic Theorem 3.5 no longer applies. For the variable case, it was shown in section 11.1.5 of [OR], that, under the same conditions as before, for any x^0 sufficiently close to x^* and any sequence $m_k \geq 1$, the iterates $\{x^k\}$ of (6.5) are well defined and converge to x^* with

$$R_1\{x^k\} \leq \rho(H(x^*))^{m'}, \quad m' = \liminf_{k\to\infty} m_k.$$

In particular, if $\lim_{k\to\infty} m_k = +\infty$, then the convergence is R-superlinear.

There is, of course, no need to restrict consideration to the "classical" linear iterative processes mentioned so far and, in fact, the methods considered in section 6.4 work instead with Krylov methods such as GMRES or the conjugate gradient method.

6.2 Nonlinear SOR Methods

This section digresses to some related results concerning a different class of primary iterations. All nonlinear iterative methods discussed so far approximated a solution of a nonlinear equation in R^n by the solutions of a sequence of n-dimensional linear equations. Instead, we may also use the solutions of a sequence of "simpler" nonlinear equations, each of which requires, of course, the use of some secondary iterative process.

The best-known methods of this type derive as natural generalizations of linear iterative processes. At the kth step of the linear Gauss–Seidel process, the component x_i^{k+1} of the next iterate is obtained by solving the ith equation with respect to its ith variable while all other variables are held at their latest values. This prescription may also be applied to a nonlinear system involving a mapping $F : E \subset \mathrm{R}^n \mapsto \mathrm{R}^n$ with components $f_1, \ldots, f_n$. It then requires setting x_i^{k+1} equal to a solution t of the scalar nonlinear equations

$$f_i(x_1^{k+1}, \ldots, x_{i-1}^{k+1}, t, x_{i+1}^k, \ldots, x_n^k) = 0, \quad i = 1, \ldots, n. \tag{6.10}$$

Combined with relaxation, we obtain the following generic step algorithm.

$\mathcal{G}$: **input:** $\{k, x^k, M_k\}$
$x^{k+1} := x^k$;
for $i = 1, 2, \ldots, n$
solve $f_i(x^{k+1} + te^i) = 0$ (approximately) for $t \in \mathrm{R}^1$;
(6.11) **if** solver failed **then return:** fail;
$x^{k+1} := x^{k+1} + \omega_k t e^i$;
endfor
return: $\{x^{k+1}, M_{k+1}\}$

The relaxation parameters ω_k are usually transmitted by the memory sets. Independent of their values we call this a step of the (nonlinear, cyclic) *SOR process*, even though some authors reserve this name for the case $\omega_k > 1$. In general, the scalar nonlinear equations (6.10) cannot be solved exactly. The allowable approximation tolerance for t is assumed to be specified in the memory set. If this tolerance is required to be zero, we speak of an *exact SOR process*. In the (nonlinear, cyclic) Jacobi process with relaxation the equations to be solved in the loop of (6.11) are $f_i(x^k + te^i) = 0$. In that case, the separate vector x^{k+1} is actually needed while in $\mathcal{G}$ we can simply overwrite x^k.

The nonlinear SOR and Jacobi processes were first discussed rigorously by Bers [26] for discrete analogues of certain mildly nonlinear elliptic boundary-value problems. Later studies of these methods are due to Schechter [226], [227], Ortega and Rockoff [194], Ortega and Rheinboldt [193], Rheinboldt [210], and Moré [175].

For the generalization to block methods we define first the following reduced mappings.

DEFINITION 6.1. *Let $F : E \subset \mathrm{R}^n \mapsto \mathrm{R}^n$ be given and suppose that $P_V : \mathrm{R}^n \mapsto V$ is the orthogonal projection onto the linear subspace $V \subset \mathrm{R}^n$ and*

$J_V : V \mapsto \mathrm{R}^n$ the corresponding injection. Then the reduction of F to V centered at a point $a \in \mathrm{R}^n$ is the mapping

$$G : \{v \in V \; : \; a + J_V v \in E\} \mapsto V, \; Gv = P_V F(a + J_V v) \tag{6.12}$$

and $Gv = P_V b$ is the corresponding reduction of the equation $Fx = b$.

Note that the domain of G may well be empty. We shall not dwell on this explicitly in this section, although, of course, it represents a point that has to be taken into account in any analysis. In analogy to the concept of a principal submatrix we introduce also the following special case of these reductions.

DEFINITION 6.2. *Let $F : E \subset \mathrm{R}^n \mapsto \mathrm{R}^n$ be given and $N_0 \subset \{1, 2, \ldots, n\}$ be a nonempty index set. Then the reduction* (6.12) *of F to $V = \mathrm{span}\,(e^j : j \in N_0)$ is the subfunction of F corresponding to N_0 centered at $a \in \mathrm{R}^n$.*

Evidently, for $a = x^{k+1}$ and $V = \mathrm{span}\,(e^i)$, the reduction of $Fx = 0$ is the equation $f_i(x^{k+1} + te^i) = 0$ in (6.11). This suggests the desired generalization. Let $\mathrm{R}^n = V_1 \oplus \cdots \oplus V_m$ be a direct sum of mutually orthogonal subspaces and denote by $P_i : \mathrm{R}^n \mapsto V_i$ and $J_i : V_i \mapsto \mathrm{R}^n$, $i = 1, 2, \ldots, n$, the induced projections and injections, respectively. Then, with the sequences $\{\omega_k\}_0^n$, $\{V_i, J_i, P_i\}_1^n$ transmitted by the memory sets, the algorithm

(6.13)

$\mathcal{G}$: **input:** $\{k, x^k, M_k\}$
  $x^{k+1} := x^k$;
  for $i = 1, 2, \ldots, n$
    solve $P_i F(x^{k+1} + J_i y^i) = 0$ (approximately) for $y^i \in V_i$;
    if solver failed **then return:** fail;
    $x^{k+1} := x^{k+1} + \omega_k J_i y_i$;
  endfor
  return: $\{x^{k+1}, M_{k+1}\}$

represents a step of the (nonlinear, cyclic) *generalized block-SOR process* with respect to the given decomposition. We shall speak of a standard block process if the decomposition is defined by $V_i = \mathrm{span}\,(e^j, j \in N_i)$, $i = 1, \ldots, m$, where $\{1, 2, \ldots, n\} = N_1 \cup \cdots \cup N_m$ is a partition into disjoint index sets. As in the linear case, the various block methods may be advantageous for certain systems such as those arising in the discretization of elliptic boundary-value problems.

Instead of using the parts $P_i F(x^{k+1} + J_i y^i) = 0$ cyclically, we may choose them according to some other selection principle. In that case, it is advantageous to renumber the iterates and to call each execution of the for-loop of (6.13) a step of the overall process. A general *free-steering block-SOR process* can then be characterized as follows.

(6.14)

$\mathcal{G}$: **input:** $\{k, x^k, M_k\}$
  choose a linear subspace $V \subset \mathrm{R}^n$, orthogonal projection
    $P : \mathrm{R}^n \mapsto V$ and injection $J : V \mapsto \mathrm{R}^n$;
  solve $PF(x^k + Jy) = 0$ (approximately) for $y \in V$;
  if solver failed **then return:** fail;
  $x^{k+1} := x^{k+1} + \omega_k Jy$;
  return: $\{x^{k+1}, M_{k+1}\}$

Usually, the subspace V is chosen from among the constituents of a fixed direct sum $\mathrm{R}^n = V_1 \oplus \cdots \oplus V_m$. For example, in the one-dimensional case $V_i =$ span (e^i), $i = 1, \ldots, n$, we may select $V = V_j$ such that $|f_j(x^k)| \geq |f_i(x^k)|$, for $i = 1, \ldots, m$; this is sometimes called the Seidel method. The general process is related to the method of functional averaging developed for integral and differential equations (see, e.g., Luchka [157]). Studies of block methods and free-steering processes for nonlinear equations include Schechter [227], and Rheinboldt [211].

Besides the SOR and Jacobi methods, other linear iterative processes may be extended to the nonlinear case as well. For example, with a decomposition $F = F_H + F_V$, a nonlinear ADI process may be defined by

$$(6.15) \qquad \begin{cases} \alpha x^{k+1/2} + F_H x^{k+1/2} = \alpha x^k - F_V x^k \\ \alpha x^{k+1} + F_V x^{k+1} = \alpha x^{k+1/2} - F_H x^{k+1/2} \end{cases} \qquad k = 0, 1, \ldots.$$

Here, one step of the iteration requires the solution of two nonlinear systems of the same dimension as the original one. Hence the method will be practical only if these systems are appropriately simple. For some discussion of this and related processes see, for example, Kellogg [147].

Clearly, the analysis of any of the methods in this section depends critically on the approach used to solve the nonlinear equations arising at each step. For this, a secondary iterative process is applied and then terminated after a suitable number of steps. Since, in principle, any type of one-dimensional process may be used here, we are led to a large variety of combined methods, and a few typical examples should suffice.

If one Newton step is applied to the equations (6.10) we obtain the (one-step) *SOR–Newton process*

$$(6.16) \qquad x_i^{k+1} = x_i^k - \omega_k \frac{f_i(x_1^{k+1}, \ldots, x_{i-1}^{k+1}, x_i^k, \ldots, x_n^k)}{\partial_1 f_i(x_1^{k+1}, \ldots, x_{i-1}^{k+1}, x_i^k, \ldots, x_n^k)}, \qquad \begin{array}{l} i = 1, \ldots, n, \\ k = 0, 1, \ldots. \end{array}$$

Analogously the (one-step) *Jacobi–Newton method* has the form

$$x_i^{k+1} = x_i^k - \omega_k \frac{f_i(x^k)}{\partial_i f_i(x^k)}, \qquad i = 1, \ldots, n, k = 0, 1, \ldots,$$

which is identical to the (one-step) Newton–Jacobi method (6.6). These processes were proposed by Lieberstein [156] and then studied by several authors; see, for example, Schechter [226], [227], Greenspan and Yohe [118], Bryan [45], Greenspan [115], and Ortega and Rockoff [194].

Newton's method may, of course, also be applied in the case of the ADI process (6.15). The resulting (one-step) *ADI–Newton method* has the form

$$\begin{cases} x^{k+1/2} = x^k - [\alpha I + DF_H(x^k)]^{-1} F x^k, \\ x^{k+1} = x^{k+1/2} - [\alpha I + DF_V(x^k)]^{-1} F x^{k+1/2}. \end{cases}$$

Instead of Newton's method as the secondary process we may use other one-dimensional iterations. For example, the (one-step) *Jacobi–Steffensen method* (see, e.g., Wegge [277]) has the form

$$x_i^{k+1} = x_i^k - \omega_k \frac{f_i(x^k)^2}{f_i(x^k) - f_i(x^k - f_i(x^k)e^i)}, \quad i = 1, \ldots, n, k = 0, 1, \ldots.$$

In general, an explicit representation of the corresponding m-step processes tends to be rather complicated. The local convergence behavior of the above combined processes involving Newton's method as the secondary iteration may be deduced from Theorem 3.5. For instance, in the case of (6.16) we obtain the following result.

THEOREM 6.1. *Let F be C^1 on an open neighborhood $\mathcal{U} \subset E$ of the simple zero $x^* \in E$. If the linear SOR iteration applied to the limiting linear system $DF(x^*)y = b$ converges for any starting point, then x^* is a point of attraction of the one-step SOR–Newton process $\mathcal{J}$ defined by* (6.16), *and the R_1-factor of both methods is the same.*

An analogous result holds for the (one-step) Jacobi–Newton method. In [OR] a corresponding theorem for a general one-step ADI–Newton process is proved, and Voigt [270] gave rate-of-convergence results for four processes combining the SOR or the Jacobi method with one step of the sequential secant or the regula falsi iteration.

For the exact SOR or Jacobi processes, local convergence may be established by means of Theorem 3.7. For instance, in the case of the Gauss–Seidel method we may define the mapping $G : E \times E \subset \mathrm{R}^n \times \mathrm{R}^n \mapsto \mathrm{R}^n$ by $g_i(y, x) = f_i(y_1, \ldots, y_i, x_{i+1}, \ldots, x_n)$, for $i = 1, \ldots, n$. Then the equations (6.10) are indeed equivalent with $G(x^{k+1}, x^k) = 0$. Straightforward differentiation now shows that

$$-\partial_1 G(x^*, x^*)^{-1} \partial_2 G(x^*, x^*) = H_1(x^*),$$

where $H_1(x^*)$ is the SOR matrix (with $\omega = 1$) for the limiting system $DF(x^*)y = b$. In this manner we obtain the following result.

COROLLARY 6.2. *Under the assumptions of Theorem 6.1, x^* is also a point of attraction of the exact SOR process with exactly the same R_1-factor.*

The nonlinear SOR iteration is the limiting form of an m-step SOR–Newton iteration as m tends to infinity. Since the one-step method and the "infinite-step" method have the same asymptotic rate of convergence, it is reasonable that the m-step process also has the same R_1-factor. Therefore, it cannot be expected that the asymptotic rate of convergence is improved by taking more than one Newton step.

6.3 Residual Convergence Controls

In this section we return to the setting of section 6.1. As indicated there, it is desirable to introduce adaptive stopping criteria for the secondary processes. The design of such adaptive controls will be based on several theorems characterizing certain conditions for the convergence of general sequences in R^n. We follow here in part Dembo, Eisenstat, and Steihaug [62], and Eisenstat and Walker [89].

If Newton's method is used as the primary process, then at each step the linear system has the form $DF(x^k)s = Fx^k$ and any secondary iteration will provide us only with an approximate solution $\hat{s}^k$. In other words, the next iterate is $x^{k+1} = x^k - \hat{s}^k$ and the residual

$$r^k = DF(x^k)(x^{k+1} - x^k) + Fx^k \tag{6.17}$$

is not expected to be zero. However, as long as the norm $\|Fx^k\|$ of the primary residual is still relatively large, there is little reason for forcing $\|r^k\|$ to be very small. This suggests that an adaptive control of the secondary iteration be based on the quotient $\|r^k\|/\|Fx^k\|$ of the secondary and primary residuals. The following theorem shows that we can expect convergence when this quotient remains bounded below one.

THEOREM 6.3. *Let $F : E \subset \mathrm{R}^n \mapsto \mathrm{R}^n$ be a C^1 mapping on the open set E which has a simple zero $x^* \in E$ where*

$$\beta = \|DF(x^*)\|, \quad \gamma = \|DF(x^*)^{-1}\|, \quad \kappa = \beta\gamma. \tag{6.18}$$

Then, for given $\eta \in (0,1)$, there exists a $\delta > 0$ such that $\mathcal{B} \equiv \bar{B}(x^*, \kappa\delta) \subset E$ and that any sequence $\{x^k\}_0^\infty \subset E$ with the properties

$$\|x^0 - x^*\| \le \delta, \tag{6.19}$$

$$\|DF(x^k)(x^{k+1} - x^k) + Fx^k\| \le \eta\|Fx^k\|, \quad \forall\, k \ge 0, \tag{6.20}$$

satisfies $x^k \in \mathcal{B}$, $\forall\, k \ge 0$, and $\lim_{k\to\infty} x^k = x^*$.

Proof. We introduce the scaled norm

$$\|y\|_* = \|DF(x^*)y\|, \quad \forall\, y \in \mathrm{R}^n, \tag{6.21}$$

for which evidently

$$\frac{1}{\gamma}\|y\| \le \|y\|_* \le \beta\|y\|, \quad \forall\, y \in \mathrm{R}^n. \tag{6.22}$$

Now with

$$\epsilon_0 = \frac{1-\eta}{\gamma(\eta+3)} < \frac{1}{\gamma} \tag{6.23}$$

we obtain

$$\alpha \equiv \alpha(\epsilon) = \frac{\eta + \gamma\epsilon(\eta+2)}{1-\gamma\epsilon} < 1, \quad \forall\, \epsilon \in (0, \epsilon_0). \tag{6.24}$$

For given $\epsilon \in (0, \epsilon_0)$ choose $\delta > 0$ such that $\mathcal{B} = \bar{B}(x^*, \kappa\delta) \subset E$ and $\|DF(x) - DF(x^*)\| \le \epsilon$ for $x \in \mathcal{B}$ whence

$$\begin{aligned} &\|Fx - Fx^* - DF(x^*)(x - x^*)\| \\ &\qquad = \|\textstyle\int_0^1 [DF(x^* + t(x - x^*)) - DF(x^*)](x - x^*)dt\| \\ &\qquad \le \epsilon\|x - x^*\|, \quad \forall\, x \in \mathcal{B}. \end{aligned} \tag{6.25}$$

From $\|DF(x)DF(x^*)^{-1} - I_n\| \leq \gamma\epsilon < 1$ it follows that $DF(x)DF(x^*)^{-1}$ is nonsingular and that

$$\|DF(x^*)DF(x)^{-1}\| \leq \mu = \frac{1}{1-\gamma\epsilon}. \tag{6.26}$$

By induction we show that for all $k \geq 0$

$$\begin{array}{lll} \text{(a)} & x^j \in \mathcal{B}, \quad \text{for } j = 0, \ldots, k, \\ \text{(b)} & \|x^{j+1} - x^*\|_* \leq \alpha\|x^j - x^*\|_*, \quad \text{for } j = 0, \ldots, k-1. \end{array} \tag{6.27}$$

For $k = 0$, (a) follows from (6.19) and (b) is vacuous. Suppose that (6.27) holds for some $k \geq 0$. Then

$$\begin{aligned} DF(x^*)(x^{k+1} - x^*) = DF(x^*)DF(x^k)^{-1}[&DF(x^k)(x^{k+1} - x^k) + Fx^k \\ &+(DF(x^k) - DF(x^*))(x^k - x^*) \\ &-(Fx^k - Fx^* - DF(x^*)(x^k - x^*))] \end{aligned}$$

together with (6.20) provides the estimate

$$\begin{aligned} \|x^{k+1} - x^*\|_* &\leq \mu\Big[\ \|r^k\| + 2\epsilon\|x^k - x^*\|\ \Big] \\ &\leq \mu\Big[\ \eta\|Fx^k\| + 2\epsilon\|x^k - x^*\|\ \Big]. \end{aligned} \tag{6.28}$$

Because

$$Fx^k = DF(x^k)(x^k - x^*) + \big[Fx^k - Fx^* - DF(x^*)(x^k - x^*)\big], \tag{6.29}$$

(6.25) implies that $\|Fx^k\| \leq \|x^k - x^*\|_* + \epsilon\|x^k - x^*\|$, and we obtain from (6.28) and (6.22)

$$\begin{aligned} \|x^{k+1} - x^*\|_* &\leq \mu\Big[\eta\|(x^k - x^*)\|_* + \epsilon(\eta + 2)\|x^k - x^*\|\Big] \\ &\leq \mu\big[\eta + \gamma\epsilon(\eta + 2)\big]\ \|x^k - x^*\|_* = \alpha\|x^k - x^*\|_*. \end{aligned} \tag{6.30}$$

Thus (6.27)(b) holds, and therefore

$$\|x^{k+1} - x^*\| \leq \gamma\alpha^{k+1}\|x^0 - x^*\|_* \leq \kappa\|x^0 - x^*\| \leq \kappa\delta \tag{6.31}$$

shows that $x^{k+1} \in \mathcal{B}$. This completes the induction and with it the proof since the convergence follows directly from (6.27)(b). □

It is straightforward to supplement this result with a rate-of-convergence statement.

COROLLARY 6.4. *Suppose that the conditions of Theorem* 6.3 *hold. Then, for any sequence* $\{x^k\}_0^\infty \subset E$ *satisfying* (6.19) *and* (6.20), *we have* $Q_1^*\{x^k\} = \eta$ *under the scaled norm* (6.21). *Moreover, the convergence is Q-superlinear exactly if*

$$\|r^k\| = \mathrm{o}\ (\|Fx^k\|),\ \ k \to \infty. \tag{6.32}$$

Proof. We use the notation and details of the proof of Theorem 6.3 and note that the estimate (4.13) of Theorem 4.2 holds here verbatim; that is

$$(6.33) \qquad (\beta+\epsilon)\|x-x^*\| \geq \|Fx\| \geq \left(\frac{1}{\gamma}-\epsilon\right)\|x-x^*\|.$$

Clearly, (6.27)(b) implies that $Q_1^*\{x^k\} \leq \alpha$ and, since $\epsilon > 0$ is arbitrarily small, it follows that $Q_1^*\{x^k\} = \alpha(0) = \eta$ as claimed. Suppose that the convergence is Q-superlinear. From

$$\begin{aligned} r^k &= Fx^k - Fx^* - DF(x^*)(x^k - x^*) - [DF(x^k) - DF(x^*)](x^k - x^*) \\ &\quad + \big[DF(x^*) + (DF(x^k) - DF(x^*))\big](x^{k+1} - x^*) \end{aligned}$$

and (6.25) we obtain $\|r^k\| \leq 2\epsilon\|x^k - x^*\| + (\beta+\epsilon)\|x^{k+1} - x^*\|$. The Q-superlinear convergence requires that $\|x^{k+1} - x^*\| = o\,(\|x^k - x^*\|)$ as $k \to \infty$ whence $\|r^k\| = o\,(\|x^k - x^*\|)$, which, by (6.33), proves (6.32). Conversely, if (6.32) holds and, thus by (6.33) also $\|r^k\| = o\,(\|x^k - x^*\|)$, then the first line of (6.28) shows that $\|x^{k+1} - x^*\| = o\,(\|x^k - x^*\|)$. □

As an application of these results, we obtain a version of Theorem 5.3 about discretized Newton methods which relates to a result of Ypma [285]. More specifically, we can prove the local convergence of the following class of "approximate" Newton methods involving a parameter $\nu > 0$.

$\mathcal{G}$: **input:**$\{k, x^k, M_k\}$
choose a nonsingular $A_k \in L(\mathrm{R}^n)$ with $\|A_k - DF(x^k)\| < \nu$;
(6.34) solve $A_k y = Fx^k$ for y;
$x^{k+1} := x^k - y$;
return: $\{x^{k+1}, M_{k+1}\}$

COROLLARY 6.5. *Let the conditions of Theorem* 6.3 *hold. Then there exists a constant* $\nu > 0$ *such that, starting from any* $x^0 \in E$ *for which* (6.19) *holds, the sequence* $\{x^k\}_0^\infty$ *generated by the algorithm* (6.34) *remains in the domain* E *and satisfies* (6.20). *Hence for this sequence the conclusions of Theorem* 6.3 *and Corollary* 6.4 *hold.*

Proof. Let ϵ_0 be given by (6.23) and for $\epsilon \in (0, \epsilon_0)$ choose $\delta > 0$ as specified in the proof of Theorem 4.4. With

$$(6.35) \qquad \nu = \frac{\eta}{1+\eta}\,\frac{1-\gamma\epsilon}{\gamma}$$

we show by induction that $x^j \in \mathcal{B}$ for $j = 0, \ldots, k$, and that (6.20) holds for $j = 0, \ldots, k-1$. For $k = 0$ this is obviously valid. If it holds for some $k \geq 0$ then by (6.26) we have

$$\|DF(x^k)^{-1}\| \leq \gamma\|DF(x^*)DF(x^k)^{-1}\| \leq \frac{\gamma}{1-\gamma\epsilon}$$

and therefore, with the choice (6.35) of ν, the perturbation lemma implies that

$$\|A_k^{-1}\| \leq \frac{\gamma}{1-\gamma(\nu+\epsilon)}.$$

Hence it follows that

$$\begin{aligned}\|DF(x^k)(x^{k+1}-x^k)+Fx^k\| &= \|-DF(x^k)A_k^{-1}Fx^k+Fx^k\|\\ &= \|(A_k-DF(x^k))A_k^{-1}Fx^k\| \le \nu\|A_k^{-1}\|\|Fx^k\|\\ &\le \frac{\nu\gamma}{1-\gamma(\nu+\epsilon)}\|Fx^k\| = \eta\|Fx^k\|.\end{aligned}$$

Now, we continue as in the proof of Theorem 6.3 and conclude that (6.28) and therefore (6.30) and (6.31) hold. This proves that $x^{k+1}\in\mathcal{B}$ and completes the induction step and with it the proof. □

If $\{x^k\}\subset E$ represents the sequence generated by Newton's method then, with $\eta=0$, Theorem 6.3 and Corollary 6.4 state that, for x^0 sufficiently close to x^*, the sequence $\{x^k\}$ converges Q- and R-superlinearly to x^*. This is equivalent to the statement of part (i) of Theorem 4.4. Part (ii) of the latter theorem asserts that close to the solution x^* the residuals $\|Fx^k\|$ converge monotonically to zero. It is interesting that this monotonicity condition together with the control condition (6.20) suffices to ensure that the sequence under consideration either has no limit point or converges to a simple zero of F.

THEOREM 6.6. *Suppose that $F: E\subset \mathrm{R}^n\mapsto\mathrm{R}^n$ is C^1 on an open set E. Let $\{x^k\}\subset E$ be any sequence such that $\lim_{k\to\infty}Fx^k=0$ and that for all $k\ge 0$*

$$\begin{aligned}&(6.36)\qquad & \|Fx^k+DF(x^k)(x^{k+1}-x^k)\| &\le \eta\|Fx^k\|,\\ &(6.37)\qquad & \|Fx^{k+1}\| &\le \|Fx^k\|,\end{aligned}$$

where $\eta>0$ is independent of k but is not otherwise restricted. If $\{x^k\}$ has any limit point $x^*\in E$ where $DF(x^*)$ is nonsingular, then $Fx^*=0$ and $\lim_{k\to\infty}x^k=x^*$.

Proof. By $\lim_{k\to\infty}Fx^k=0$ and the continuity of F, any limit point $x^*\in E$ of $\{x^k\}$ is a zero of F. Suppose that x^* is a simple zero where we use again the notation (6.18). Choose $\delta>0$ such that $\mathcal{B}=\bar{B}(x^*,\kappa\delta)\subset E$ and $\|DF(x)-DF(x^*)\|\le 1/2\gamma$. Then, by the perturbation lemma, $DF(x)$ is invertible with $\|DF(x)^{-1}\|\le 2\gamma$ for all $x\in\mathcal{B}$, and as in (6.25) we conclude that

$$\|Fx-Fx^*-DF(x^*)(x-x^*)\|\le\frac{1}{2\gamma}\|x-x^*\|,\quad x\in\mathcal{B}.$$

Thus from (6.29) with x in place of x^k, it follows that

$$\|Fx\|\ge\frac{1}{\gamma}\|x-x^*\|-\frac{1}{2\gamma}\|x-x^*\|=\frac{1}{2\gamma}\|x-x^*\|$$

and therefore that

$$(6.38)\qquad \|x-x^*\|\le 2\gamma\|Fx\|,\quad \forall\, x\in B(x^*,\delta).$$

Let $\epsilon\in(0,\delta/4)$. Clearly, there exists a sufficiently large $k\ge 0$ such that

$$(6.39)\qquad \text{(a)}\quad \|Fx^k\|<\frac{\epsilon}{\gamma(1+\eta)},\qquad \text{(b)}\quad \|x^k-x^*\|<\frac{2\gamma\epsilon}{\gamma(1+\eta)}<\frac{\delta}{2}.$$

Then

$$\begin{aligned}&\|x^{k+1} - x^k\|\\ &\quad \leq \|DF(x^k)^{-1}\| \ \left[\|Fx^k + DF(x^k)(x^{k+1} - x^k)\| \ + \ \|Fx^k\|\right]\\ &\quad \leq 2\gamma(1+\eta) \ \|Fx^k\| < 2\epsilon < \frac{\delta}{2},\end{aligned}$$

whence, with (6.39)(b), $\|x^{k+1} - x^*\| < \delta$; that is, $x^{k+1} \in B(x^*, \delta)$. Now, by (6.37) we have

$$\|Fx^{k+1}\| \leq \|Fx^k\| < \frac{\epsilon}{\gamma(\eta+1)}$$

and, with (6.38),

$$\|x^{k+1} - x^*\| \leq 2\gamma\|Fx^{k+1}\| < \frac{2\gamma\epsilon}{\gamma(1+\eta)} < \frac{\delta}{2};$$

that is, the two conditions (a) and (b) of (6.39) are satisfied for x^{k+1} in place of x^k. Therefore, by induction it follows that for $j \geq k$ the points x^j satisfy (6.39) with j in place of k. In particular, we see that $x^j \in B(x^*, \delta)$ for $j \geq k$ and hence, because of $Fx^j \to 0$, that $x^j \to x^*$. □

6.4 Inexact Newton Methods

The theorems of the previous section are nonconstructive in nature. This section shows how to develop algorithms from these results. In particular, we consider a class of combination processes with Newton's method as the primary iteration, which carry the name *inexact Newton methods* given to them by Dembo, Eisenstat, and Steihaug [62].

In preparation we show first that when the monotonicity condition (6.37) of Theorem 6.6 is strengthened then we can drop the requirement that the residuals converge to zero. For this it will be assumed that F is defined on all of R^n.

THEOREM 6.7. *Given a C^1 map $F : \mathrm{R}^n \mapsto \mathrm{R}^n$, let $\{x^k\} \subset \mathrm{R}^n$ be any sequence such that for all $k \geq 0$*

$$\|DF(x^k)(x^{k+1} - x^k) + Fx^k\| \leq \eta_k\|Fx^k\|, \ 0 < \eta_k \leq \eta < 1, \tag{6.40}$$

$$\|F(x^{k+1})\| \leq \|Fx^k\|, 0 < \lambda_k \leq \lambda < 1. \tag{6.41}$$

If the sequence has a limit point x^* where $DF(x^*)$ is invertible, then $\lim_{k\to\infty} x^k = x^*$ *and* $Fx^* = 0$.

Proof. For any convergent subsequence $\{x^{k_j}\}$ it follows from (6.41) that $\lim_{j\to\infty} Fx^{k_j} = 0$ whence, by continuity of F, the limit point must be a zero of F. In particular, x^* is a simple zero of F. With the notation (6.18) and as in the proof of Theorem 6.6, let $\delta > 0$ be such that $\|DF(x)^{-1}\| \leq 2\gamma$ for $x \in \mathcal{B} = \bar{B}(x^*, \delta)$. If $x^k \in \mathcal{B}$ for some index k then it follows from (6.40) that

$$\begin{aligned}\|x^{k+1} - x^k\| &\leq 2\gamma\left[\|-Fx^k\| + \|Fx^k + DF(x^k)(x^{k+1} - x^k)\|\right]\\ &\leq 2\gamma(1+\eta_k)\|Fx^k\| \leq 4\gamma\|Fx^k\|.\end{aligned} \tag{6.42}$$

Suppose now that the subsequence $\{x^{k_j}\}$ converges to x^* but that the entire sequence does not converge to that point. By reducing δ if needed, we know from Theorem 4.2 that x^* is the only zero of F in $\mathcal{B}$. Thus we must have $x^k \notin B(x^*, \delta)$ for infinitely many k. It is no restriction to assume that $\{x^{k_j}\} \subset B(x^*, \delta/2)$ and that for each k_j there exists an index ℓ_j, $0 < \ell_j < k_{j+1} - k_j$ for which

$$x^{k_j+i} \in B(x^*, \delta),\ i = 0, \ldots, \ell_j - 1, x^{k_j+\ell_j} \notin B(x^*, \delta).$$

Then (6.42) and (6.41) imply that

$$\|x^{i+1} - x^i\| \leq 4\gamma\|Fx^i\| \leq \frac{4\gamma}{1-\lambda}[\|Fx^i\| - \|Fx^{i+1}\|], i = k_j, \ldots, k_j + \ell_j - 1.$$

Hence we obtain

$$\begin{aligned} \frac{\delta}{2} &\leq \|x^{k_j+\ell_j} - x^{k_j}\| \leq \sum_{i=k_j}^{k_j+\ell_j-1} \|x^{i+1} - x^i\|, \\ &\leq \sum_{i=k_j}^{k_j+\ell_j-1} \frac{4\gamma}{1-\lambda}\left[\|Fx^i\| - \|Fx^{i+1}\|\right] \\ &= \frac{4\gamma}{1-\lambda}[\|F(x^{k_j})\| - \|F(x^{k_j+\ell_j})\|] \\ &\leq \frac{4\gamma}{1-\lambda}[\|Fx^{k_j}\| - \|Fx^{k_j+1}\|], \end{aligned}$$

where in the last inequality we again used the monotonicity assumption (6.41). This provides a contradiction, since as we saw, $\lim_{j\to\infty} Fx^{k_j} = 0$. Therefore the entire sequence $\{x^k\}$ must converge to x^* and, as noted already, x^* is a simple zero of F. □

The question is now how to guarantee the validity of (6.40) and (6.41) at each step. This can be accomplished by a step reduction. For a given point $x \in \mathbb{R}^n$, $Fx \neq 0$, and any step $s \in \mathbb{R}^n$ we consider the two control variables

(a) residual quotient: $$r(s) = \frac{\|Fx + DF(x)s\|}{\|Fx\|},$$

(b) relative step: $$\sigma(s) = \frac{\|s\|}{\|Fx\|}.$$

When a step $\hat{s}$ has been found such that $r(\hat{s}) < 1$ then for any reduced step $s = \theta\hat{s}$, $\theta \in (0, 1)$, it follows easily that

$$r(s) \leq (1-\theta) + \theta r(\hat{s}) < 1, \quad \sigma(s) = \theta\sigma(\hat{s}). \tag{6.43}$$

For the computation it is no restriction to assume that the entire process remains in a compact, convex set $C \subset \mathbb{R}^n$. Then, by uniform continuity, for any small $\epsilon > 0$ there exists $\rho = \rho(\epsilon)$ such that $\|DF(y) - DF(x)\| \leq \epsilon$ whenever $\|y - x\| \leq \rho$

for $x, y \in C$. Hence, under these conditions for x, y, we have

$$
\begin{aligned}
\|Fy &- Fx - DF(x)(y-x)\| \\
&\le \left\| \int_0^1 [DF(x+t(y-x)) - DF(x)](y-x)dt \right\| \qquad (6.44)\\
&\le \frac{1}{2}\epsilon\|y-x\|,
\end{aligned}
$$

and for $x, x+s \in C, \|s\| \le \rho$ it follows that

$$
\begin{aligned}
\|F(x+s)\| &\le \|Fx + DF(x)s\| + \|F(x+s) - Fx - DF(x)s\| \\
&\le r(s)\|Fx\| + \frac{1}{2}\epsilon\|s\| \qquad (6.45)\\
&\le \mu(s)\|Fx\|, \quad \mu(s) = r(s) + \frac{1}{2}\epsilon\sigma(s).
\end{aligned}
$$

Thus, for given x and $\hat{s}$ such that $x, x+\hat{s} \in C$ we can choose ϵ such that $\mu(\hat{s}) = r(\hat{s}) + \frac{1}{2}\epsilon\sigma(\hat{s}) < 1$. Then, for small enough $\theta \in (0,1)$ such that $\|s\| = \theta\|\hat{s}\| < \rho(\epsilon)$, it follows from (6.45) that $\|F(x+s)\| \le \mu(s)\|Fx\|$ with $\mu(s) \le (1-\theta) + \theta\mu(\hat{s}) < 1$. This shows that with a step reduction we can satisfy both conditions (6.40) and (6.41) if only the first one already holds for the initial step $\hat{s}$.

This suggests the following "minimum reduction" algorithm of Eisenstat and Walker [89]. It involves the choice of suitable parameters $\tau \in (0,1)$, $\eta_{\max}$, $0 < \theta_{\min} < \theta_{\max} < 1$, and $j_{\max}$, which here are assumed to be supplied via the memory sets. A typical acceptance test for methods of this type uses a residual condition $\|Fx^k\| \le \text{tol}$ with an appropriate tolerance.

(6.46)

$\mathcal{G}$: **input:** $\{k, x^k, M_k\}$
determine $\hat{s}^k$, and $\hat{\eta}_k \in (0, \eta_{\max})$
  such that $\|Fx^k + DF(x^k)\hat{s}^k\| \le \hat{\eta}_k\|Fx^k\|$;
$\eta := \hat{\eta}_k$; $s := \hat{s}^k$; $j := 0$;
while $\|F(x^k+s)\| > [1-\tau(1-\eta)]\ \|Fx^k\| \ \bigwedge\ j \le j_{\max}$
  choose $\theta \in [\theta_{\min}, \theta_{\max}]$;
  $\eta := (1-\theta) + \theta\eta$; $s := \theta s$;
endwhile
if $j > j_{\max}$ **then return:** fail;
$x^{k+1} := x^k + s$;
return: $\{x^{k+1}, M_{k+1}\}$

The while loop is executed only when $r(\hat{s}^k) < 1$. Thus our earlier observations apply. For sufficiently small $\theta_{\min}$ we may expect that the condition $\|s\| \le \rho(\epsilon)$ can be reached. Since $1-\eta$ is reduced by a factor $\theta \le \theta_{\max} < 1$ each time the while loop is repeated, this loop should terminate with $j \le j_{\max}$ provided $j_{\max}$ was not chosen too small.

When there is no failure we obtain the following result.

THEOREM 6.8. *Suppose that the algorithm* (6.46) *does not fail and that* x^* *is a limit point of the computed sequence* $\{x^k\}$ *where* $DF(x^*)$ *is invertible. Then* $Fx^* = 0$ *and* $\lim_{k\to\infty} x^k = x^*$. *Moreover, we have* $\eta_k = \hat{\eta}_k$ *for all sufficiently large* k.

Proof. When the algorithm does not fail, then the computed sequence satisfies both (6.40) and (6.41). Hence the convergence follows from Theorem 6.7. For the proof of the last statement let $\delta > 0$ be as specified in the proof of Theorem 6.7. The iterates are certainly contained in some compact, convex set $C \subset \mathbb{R}^n$ for which there exists $\rho = \rho(\epsilon)$ such that $\|DF(y) - DF(x)\| \leq \epsilon$ whenever $x, y \in C$ and $\|y - x\| \leq \rho$. It is no restriction to assume that $\delta \leq \rho(\epsilon)$ for $\epsilon = (1-\tau)(1-\eta_{\max})/(4\gamma)$ where $\tau \in (0,1)$ is the constant used in (6.46). For sufficiently large k we have $x^k \in \mathcal{B} = \bar{B}(x^*, \delta)$ and $\|Fx^k\| \leq \delta/(4\gamma)$. Hence, as in the proof of (6.42) in Theorem 6.7, it follows from (6.40) that $\|\hat{s}^k\| \leq 4\gamma\|Fx^k\| \leq \delta \leq \rho$. Therefore (6.45) applies and we obtain $\|Fx^{k+1}\| \leq \mu(\hat{s}^k)\|Fx^k\|$ with

$$\mu(\hat{s}^k) \leq r(\hat{s}^k) + \epsilon\sigma(\hat{s}^k) \leq \eta_{\max} + (1-\tau)\frac{1-\eta_{\max}}{4\gamma}4\gamma = 1 - \tau(1-\eta_{\max}) < 1.$$

In other words, (6.41) holds already for $\hat{s}^k$ and no step reduction is needed. □

The algorithm (6.46) now requires at each step a secondary iterative method for computing an approximate solution $\hat{s}$ of the linear system $DF(x^k)s + Fx^k = 0$ which meets the condition in the algorithm (6.46) just before entering the whileloop. In principle, of course, any linear iterative process could be applied, but, as proposed by Brown and Saad [38], Turner and Walker [268], Walker [274], and Eisenstat and Walker [89], [88], the general minimum residual method GMRES of Saad and Schultz [218] has found widespread use.

We refer to Saad and Schultz [218], Walker [272], [273], or Kelley [146] for details about the GMRES iteration for solving a linear system $Ax = b$. Briefly, the kth iterate x^k of GMRES is the solution of the least-squares problem

$$\min_{x \in x^0 + \mathcal{K}_k} \|b - Ax\|_2,$$

where $\mathcal{K}_k = \text{span}\,\{r^0, Ar^0, \ldots, A^{k-1}r^0\}$ is the kth Krylov subspace generated by A and the residual $r^0 = b - Ax^0$ at the starting point x^0. If A is nonsingular then (in real arithmetic) the process terminates at the solution after at most n steps, but for the computation the method is best set up as an iterative method which stops only when a convergence criterion, such as $\|r^k\|_2 \leq \epsilon\|r^0\|_2$, is met. The solution of the least-squares problem follows standard procedures once orthonormal bases of the spaces $\mathcal{K}_k$ have been computed. For this, as proposed by Arnoldi [11], a modified Gram–Schmidt procedure can be applied effectively. Then a basic version of GMRES has the form (see, e.g., Kelley [146])

(6.47)

```
GMRES: input: {x, b, A, ε, m, p}
       for ℓ = 1, ..., p do
         r := b − Ax; ρ := ‖r‖₂; u¹ := r/ρ;
         for k = 1, ..., m do
           u^{k+1} := Au^k;
           for j = 1, ..., k do
             h_{jk} := (u^{k+1})ᵀ u^j; u^{k+1} := u^{k+1} − h_{jk} u^j;
           endfor
           h_{k+1,k} := ‖u^{k+1}‖₂; u^{k+1} := u^{k+1}/h_{k+1,k};
           H := (h_{ji}) ∈ L(R^k, R^{k+1}) with h_{ji} = 0, j > i − 1;
           y = min_{z∈R^{k+1}} ‖ρe¹ − Hz‖₂ with e¹ ∈ R^{k+1};
           if ‖ρe¹ − Hy‖₂ ≤ ε‖b‖₂ then
             return: x := x + (u¹, ..., u^{k+1})y;
           endif
         endfor
         x := x + (u¹, ..., u^{m+1})y;
       endfor
       return error
```

Since the costs of storage and arithmetic increase with the step count k, only a maximum number m of steps is allowed and, if after the execution of a cycle of m steps the process has not terminated then a new cycle of m steps is started up to a maximum of p cycles. Instead of the modified Gram–Schmidt procedure used here, it is also possible to use Given's rotations (see Saad and Schultz [218]), or Householder reflections (see Walker [273]). Various implementations of Newton-GMRES algorithms have been given in the cited literature.

Among other advances related to the topic of this section we mention only the inexact Gauss–Newton methods considered by Hohmann [129] and the inexact Newton methods with a preconditioned conjugate gradient iteration as secondary process for the finite-element solution of elliptic problems developed by Deuflhard and Weiser [75]. The various forms of the inexact Newton methods are at present widely considered to be among the best approaches for solving nonlinear systems of equations.

CHAPTER 7

Parametrized Systems of Equations

7.1 Submanifolds of $\mathbf{R}^n$

Equilibrium problems for many physical systems are modeled by parameter-dependent equations of the form

$$F(z, \lambda) = 0, \tag{7.1}$$

where z denotes some state variable and λ a vector of parameters. An example is Bratu's problem (2.1) and its discretizations discussed in section 2.1. In practice it is rarely sufficient to compute the solution z of (7.1) for a few chosen values of λ. Instead interest centers on understanding the variation of the states with changing parameters, and, in particular, on determining those parameter values where the character of the state changes, as for instance where a mechanical structure buckles. These are exactly the situations where the state can no longer be expressed as a continuous function of λ; that is, where we need to consider the general solution set

$$M = \{(z, \lambda) \ : \ F(z, \lambda) = 0\}. \tag{7.2}$$

In many cases, M is a differentiable manifold. In the engineering literature this is recognized by the terminology "equilibrium surface" for the set M, although often no mathematical characterization of the manifold structure is provided.

A C^1 map $F : E \subset \mathrm{R}^n \mapsto \mathrm{R}^m$ on an open set E is an *immersion* or *submersion* at a point $x \in E$ if its derivative $DF(x) \in L(\mathrm{R}^n, \mathrm{R}^m)$ is a one-to-one (linear) mapping or a mapping onto R^m, respectively. We call F a submersion or immersion on a subset $E_0 \subset E$ if it has that property at each point of E_0. These definitions require $n \leq m$ for F to be an immersion and $n \geq m$ for it to be a submersion. Clearly, if $n \geq m$ and $DF(x)$ has maximal rank m, then F is a submersion at x.

For an introduction to manifolds and differential geometry we refer, e.g., to Spivak [254] or Abraham, Marsden, and Ratiu [1]. Without entering into the definition of manifolds we use here simply the following characterization of submanifolds.

THEOREM 7.1. *A nonempty subset $M \subset \mathrm{R}^n$ is a submanifold of* R^n *of dimension d and class C^r, $r \geq 1$, (or a d-dimensional C^r-submanifold, for* *short), exactly if for every point* $x^c \in M$ *there exists an open neighborhood* $\mathcal{V}^n$ *of* x^c *in* R^n *and a submersion* $F : \mathcal{V}^n \mapsto \mathrm{R}^{n-d}$ *of class* C^r *such that* $M \cap \mathcal{V}^n = \{x \in \mathcal{V}^n \ : \ Fx = 0\}$.

The following special case will be used frequently.

COROLLARY 7.2. *Let* $F : E \subset \mathrm{R}^n \mapsto \mathrm{R}^m$, $n - m = d > 0$, *be of class* C^r, $r \geq 1$, *on an open subset* E *of* R^n *and a submersion on* $M = \{x \in E \ : \ Fx = 0\}$. *Then* M *is either empty or a* d*-dimensional* C^r*-submanifold of* R^n.

From Theorem 7.1 it follows readily that any nonempty, (relatively) open subset of a d-dimensional C^r-submanifold of R^n is itself a C^r-submanifold of R^n of the same dimension. For the analysis of submanifolds of R^n we need local parametrizations.

DEFINITION 7.1. *Let* $M \subset \mathrm{R}^n$ *be a nonempty set. A local* d*-dimensional* C^r*-parametrization of* M *is a pair* $(\mathcal{V}^d, \varphi)$ *consisting of a nonempty, open subset* $\mathcal{V}^d$ *of* R^d *and a* C^r*-mapping* $\varphi : \mathcal{V}^d \mapsto \mathrm{R}^n$ *such that*

(i) $\varphi(\mathcal{V}^d)$ *is (relatively) open in* M,

(ii) φ *is a homeomorphism of* $\mathcal{V}^d$ *onto its image* $\varphi(\mathcal{V}^d)$,

(iii) φ *is an immersion on* $\mathcal{V}^d$.

For any point x^c *of* M *such that* $x^c \in \varphi(\mathcal{V}^d)$ *we call* $(\mathcal{V}^d, \varphi)$ *a local* d*-dimensional* C^r*-parametrization of* M *near* x^c.

The nonempty subsets of R^n which possess a local parametrization near every point are exactly the submanifolds of R^n.

THEOREM 7.3. *A nonempty subset* $M \subset \mathrm{R}^n$ *is a* d*-dimensional* C^r*-submanifold of* R^n *if and only if for every point* $x^c \in M$ *there exists a local* d*-dimensional* C^r*-parametrization of* M *near* x^c. *If* M *is a* d*-dimensional* C^r*-submanifold of* R^n, *and* $(\mathcal{V}^{\hat{d}}, \varphi)$ *a local* $\hat{d}$*-dimensional* C^r*-parametrization of* M, *then* $\hat{d} = d$.

In line with our simplified introduction to submanifolds of R^n we will not define tangent spaces in general, and instead use the following characterization. Let M be a d-dimensional C^r-submanifold of R^n. For any point $x^c \in M$ there exists an open neighborhood $\mathcal{V}^n \subset \mathrm{R}^n$ of x^c and a submersion $F : \mathcal{V}^n \mapsto \mathrm{R}^{n-d}$ at x^c such that $M \cap \mathcal{V}^n = \{x \in \mathcal{V}^n \ : \ Fx = 0\}$. Then, it can be shown that the d-dimensional linear subspace $T = \ker DF(x^c)$ of R^n is independent of the particular choice of the local submersion F; that is, T depends only on M and the particular point. This space T is the *tangent space* of M at x^c and is denoted by $T_{x^c}M$. The subset $TM = \bigcup_{x \in M}[\{x\} \times T_xM]$ of $\mathrm{R}^n \times \mathrm{R}^n$ is the *tangent bundle* of M. As an example, we note that any open subset $E \in \mathrm{R}^n$ is an n-dimensional C^∞-submanifold of R^n for which the tangent bundle is $TE = E \times \mathrm{R}^n$. In particular, the tangent bundle of R^n is $T\mathrm{R}^n = \mathrm{R}^n \times \mathrm{R}^n$.

Thus, the tangent bundle of a submanifold M of R^n appears as a subset of the tangent bundle $T\mathrm{R}^n$, and, in general, it is a submanifold of $T\mathrm{R}^n$. In fact, if M is a d-dimensional C^r-submanifold of R^n where $r \geq 2$, then TM is a $2d$-dimensional C^{k-1}-submanifold of $T\mathrm{R}^n$. In that case, parametrizations of the C^{k-1}-submanifold TM of R^{2n} can be constructed easily from local C^r-parametrizations of M. In fact, if $(x^c, v^c) \in TM$ and $(\mathcal{V}^d, \varphi)$ is a local C^r-parametrization of M near x^c, then the pair $(\mathcal{V}^d \times \mathrm{R}^d, \ (\varphi, D\varphi))$ is a local C^{r-1}-parametrization of TM near (x^c, v^c).

The following result provides a basis for the computation of a local parametrization on submanifolds of $\mathbb{R}^n$ (see Rheinboldt [215]).

THEOREM 7.4. *Let $F : E \mapsto \mathbb{R}^m$ be of class C^r, $r \geq 1$ on an open subset E of $\mathbb{R}^n$, $n - m = d > 0$ and a submersion on $M = \{x \in E \ : \ Fx = 0\}$. Assume that $U \in L(\mathbb{R}^d, \mathbb{R}^n)$ maps $\mathbb{R}^d$ isomorphically onto a d-dimensional linear subspace $T \subset \mathbb{R}^n$, and denote by $J : \mathbb{R}^d \mapsto \mathbb{R}^m \times \mathbb{R}^d$ the canonical injection that maps $\mathbb{R}^d$ isomorphically onto $\{0\} \times \mathbb{R}^d$. Then the C^r mapping*

$$(7.3) \qquad H : E \mapsto \mathbb{R}^m \times \mathbb{R}^d, \quad Hx = (Fx, U^\top x), \ \forall \ x \in E$$

is a local diffeomorphism on an open neighborhood of $x^c \in M$ in $\mathbb{R}^n$ if and only if

$$(7.4) \qquad T_{x^c}M \cap T^\perp = \{0\}.$$

If (7.4) holds at x^c then there exists an open set $\mathcal{V}^d$ of $\mathbb{R}^d$ such that the pair $(\mathcal{V}^d, \varphi)$, defined with the mapping $\varphi = H^{-1} \circ J : \mathcal{V}^d \mapsto \mathbb{R}^n$, is a local C^r-parametrization of M near x^c.

Proof. Obviously, H is of class C^r on E. If $DH(x^c)h = 0$ for some $h \in \mathbb{R}^n$, then $DF(x^c)h = 0$, and $U^\top h = 0$, whence $h \in T_{x^c}M$, and because

$$(7.5) \qquad \langle h, Uy \rangle = \langle U^\top h, y \rangle = 0, \quad y \in \mathbb{R}^d,$$

$h \in T^\perp$. By (7.4) this implies that $h = 0$. Conversely, for any nonzero $h \in T_{x^c}M \cap T^\perp$ we have $DF(x^c)h = 0$, and (7.5) requires that $U^\top h = 0$, which together implies that $DH(x^c)h = 0$. Hence, if (7.4) holds, then there is some open neighborhood $\mathcal{V}^n$ of x^c in $\mathbb{R}^n$ such that H is a diffeomorphism from $\mathcal{V}^n$ onto the open set $H(\mathcal{V}^n)$ in $\mathbb{R}^n$. Evidently, the set $H(M \cap \mathcal{V}^n) = H(\mathcal{V}^n) \cap (\{0\} \times \mathbb{R}^d)$ is open in $\{0\} \times \mathbb{R}^d$ and $\mathcal{V}^d = J^{-1}H(M \cap \mathcal{V}^n)$ is an open subset of $\mathbb{R}^d$. This shows that $\varphi = H^{-1} \circ J$ is a C^r mapping from $\mathcal{V}^d$ onto the open subset $M \cap \mathcal{V}^n$ of M. Both φ and its inverse $J^{-1} \circ H_{|M \cap \mathcal{V}^n}$ are continuous, and hence φ is a homeomorphism of $\mathcal{V}^d$ onto $M \cap \mathcal{V}^n$. Since both H^{-1} and J are immersions, the same is true for φ. Thus, altogether, φ is a local d-dimensional C^r-parametrization of M near x^c. □

Note that (7.4) is equivalent to the conditions $\ker DF(x^c) \cap T^\perp = \{0\}$ and $\operatorname{rge} DF(x^c)^\top \cap T = \{0\}$. We call a d-dimensional linear subspace $T \subset \mathbb{R}^n$ a *coordinate subspace* of M at $x^c \in M$ if (7.4) holds. At any point $x^c \in M$ an obvious choice for a coordinate subspace is $T = T_{x^c}M$, which we will sometimes identify as the *tangential coordinate space* of M at that point.

Theorem 7.4 readily becomes a computational procedure for local parametrizations by the introduction of bases. On $\mathbb{R}^n$ and $\mathbb{R}^d$ the canonical bases will be used and we assume that the vectors $u^1, \ldots, u^d \in \mathbb{R}^n$ form an orthonormal basis of the given coordinate subspace T of M at x^c. Then the matrix representation of the mapping U is the $n \times d$ matrix with the vectors $u^1, \ldots, u^d$ as columns. We denote this matrix again by U. It is advantageous to shift the open set $\mathcal{V}^d$ such that $\varphi(0) = x^c$. Now, in component form, the nonlinear mapping H of (7.3) assumes the form

$$(7.6) \qquad H : \mathbb{R}^n \mapsto \mathbb{R}^n, \ Hx = \begin{pmatrix} Fx \\ U^\top(x - x^c) \end{pmatrix}, \quad \forall \ x \in E \subset \mathbb{R}^n.$$

By definition of φ we have

$$H(\varphi(y)) = Jy, \quad \forall\, y \in \mathcal{V}^d. \tag{7.7}$$

Thus, the evaluation of $x = \varphi(y)$ for given $y \in \mathcal{V}^d$ requires finding zeros of the nonlinear mapping

$$H_y(x) = Hx - Jy \equiv \begin{pmatrix} Fx \\ U^\top(x - x^c) - y \end{pmatrix}, \quad \forall\, x \in E. \tag{7.8}$$

Since (7.4) is assumed to hold at $x^c \in M$, the Jacobian

$$DH_y(x) = DH(x) = \begin{pmatrix} DFx \\ U^\top \end{pmatrix} \tag{7.9}$$

is nonsingular in an open neighborhood of $x = x^c \in M$. For the determination of a zero of (7.8) a chord Newton method

$$\begin{cases} x^{k+1} = N(x^k), \quad k = 0, 1, \ldots, \\ N(x) := x - A^{-1}H_y(x), \quad A = DH(x^c) \end{cases} \tag{7.10}$$

works well in practice. Recall that H (and hence also H_y) is a diffeomorphism from an open neighborhood $\mathcal{V}^n \subset E$ of x^c onto its image. Let $\epsilon \in (0, 1)$ be such that $\|A^{-1}\|\epsilon \le 1/2$. Then there exists a $\delta > 0$ such that the closed ball $\mathcal{B} = \bar{B}(x^c, \delta)$ is contained in $\mathcal{V}^n$ and that $\|DF(x) - DF(x^c)\|_2 \le \epsilon$ for all $x \in \mathcal{B}$. Hence

$$\|DN(x)\|_2 \le \|A^{-1}\|\,\|DF(x) - DF(x^c)\|_2 \le \frac{1}{2}, \quad \forall\, x \in \mathcal{B}$$

implies that N is contractive on $\mathcal{B}$. Moreover, if $\|y\|_2 \le \delta/2$ then for $x \in \mathcal{B}$ it follows from

$$\begin{aligned} \|N(x) - x^c\|_2 &\le \|N(x) - N(x^c)\|_2 + \|N(x^c) - x^c\|_2 \\ &\le \frac{1}{2}\|x - x^c\|_2 + \|y\|_2 \le \delta \end{aligned}$$

that N maps $\mathcal{B}$ into itself. Hence by the contraction-mapping Theorem 4.1, for $\|y\|_2 \le \delta/2$, the process converges for any $x^0 \in \mathcal{B}$ to the unique fixed point $x^* \in \mathcal{B} \subset \mathcal{V}^n$ of N whence $H_y(x^*) = 0$ and thus $x^* = \phi(y)$.

For the special choice $x^0 = x^c + Uy$, $\|y\|_2 \le \delta/2$, the iterates satisfy $0 = U^\top(x^k - x^c) - y = U^\top(x^k - x_0)$ whence

$$H_y(x^k) = \begin{pmatrix} Fx^k \\ 0 \end{pmatrix} \quad \forall\, k \ge 0.$$

This suggests applying the process in the form

$$x^{k+1} = x^k - A^{-1}\begin{pmatrix} Fx^k \\ 0 \end{pmatrix}, \quad x^0 = x^c + Uy, \tag{7.11}$$

where the y-dependence occurs only in the starting point. This shows that, for any local vector y near the origin of R^d, the following algorithm produces the point $x = \varphi(y)$ in the local parametrization $(\mathcal{V}^d, \varphi)$ near x^c defined by Theorem 7.4.

(7.12)

GPHI: **input:** { x^c, y, U, $DF(x^c)$, tolerances }
$x := x^c + Uy$;
compute LU factorization of $DH(x^c)$;
while iterates do not meet tolerances
evaluate Fx;
$q := \begin{pmatrix} Fx \\ 0 \end{pmatrix}$;
solve $DH(x^c)w = q$ for $w \in \mathrm{R}^n$;
$x := x - w$;
endwhile
return: $\varphi(y) := x$

In order to meet the condition (7.4) for the coordinate subspace T of M at x^c, it is useful to choose first the orthogonal complement $T^\perp$ as a complementary subspace of the tangent space $T_{x^c}M$, and then to construct T from it. Suppose that a basis $z^1, \ldots, z^m$ of $T^\perp$ has already been constructed. Then T is the nullspace of the $m \times n$ matrix Z of rank m formed with these vectors as its rows. There are several approaches for the computation of the nullspace of Z. A simple one is based on the LQ-factorization (with row pivoting)

$$Z = P^\top (L \;\; 0)\, Q^\top, \quad Q = (Q_1 \;\; Q_2). \tag{7.13}$$

Here P is an $m \times m$ permutation matrix, L an $m \times m$, nonsingular, lower-triangular matrix, and Q an $n \times n$ orthogonal matrix partitioned such that Q_1 and Q_2 are $n \times m$ and $n \times d$ matrices, respectively. Then, clearly, the d columns of Q_2 form the desired orthonormal basis of T. This justifies the following algorithm.

(7.14)

COBAS: **input:** Z
compute LQ-factorization of Z using row pivoting;
for $j = 1, 2, \ldots, d$ **do** $u^j := Q_2 e^j$;
return: $U := (u^1, \ldots, u^d)$

Other algorithms for the computation of nullspace bases of $m \times n$ matrices were given, e.g., by Berry et al. [25], and Coleman and Pothen [54].

In general, the matrix Z used in (7.14) is constructed from the Jacobian matrix $DF(x^c)$. For example, suppose we need the coordinate space T to contain, say, the ith natural basis vector e^i of R^n. This is ensured when Z is formed by zeroing the ith column of $DF(x^c)$, provided, of course, the resulting matrix still has rank m so that (7.4) holds. Obviously, when the tangential coordinate system is used at x^c, then (7.14) can be applied directly to $DF(x^c)$ as the matrix Z. In that case, (7.12) simplifies considerably if the LQ-factorization (7.13) of $Z = DF(x^c)$ is used to solve the corrector equation $DH(x^c)w = q$. In fact, this equation has the block components $DF(x)w = Fx$ and $U^\top w = 0$, which by (7.13) can be rewritten as $LQ_1^\top w = PFx$ and $Q_2 w = 0$. Thus, in this case, the algorithm can be modified as follows.

TPHI: **input:** { x^c, y, tolerances }
compute LQ-factorization of $DF(x^c)$;
$x := x^c + Q_2 y$;
while iterates do not meet tolerances
(7.15) evaluate Fx;
solve $Lv = PFx$ for $v \in \mathbb{R}^m$;
$x := x - Q_1 v$;
endwhile
return: $\varphi(y) := x$

Thus, for each iteration step we need to solve now only an $m \times m$, rather than an $n \times n$, lower-triangular system. The convergence behavior, of course, remains the same.

7.2 Continuation Using ODEs

Continuation methods are probably the oldest methods for the computational analysis of an implicitly defined manifold M. They apply to problems of the form (7.1) with a parameter space of dimension $d = 1$ and, hence, for which the solution set is expected to be a one-dimensional submanifold of $\mathbb{R}^n$.

We refer to the surveys of Allgower and Georg [4], [5] for a general overview of continuation methods and also for some historical remarks. The one-dimensional nature of the solution manifold allows for several different computational approaches. In this section we consider an approach based on the introduction of an ODE on the domain of the mapping F.

For the proof of the main result we require the following lemma proved by Rheinboldt [212].

LEMMA 7.5. *Let $A_1, A_2 \in L(\mathbb{R}^n, \mathbb{R}^{n-1})$, $n \geq 2$, both with rank $n-1$. If $u^i \in \ker A_i$, $\|u^i\|_2 = 1$, $i = 1, 2$ are such that $\alpha = (u^1)^\top u^2 \neq -1$, then*

$$\|u^1 - u^2\|_2 \leq \frac{2}{1+\alpha}\left[\min_{i=1,2} \|A_i^+\|_2\right] \|A_1 - A_2\|_2,$$

where $A_i^+ = A_i^\top (A_i A_i^\top)^{-1}$ is the pseudoinverse of A_i, $i = 1, 2$.

Proof. For $i = 1, 2$, the orthogonal projections $P_i = A_i^+ A_i$ of $\mathbb{R}^n$ onto $(\ker A_i)^\perp$ satisfy $I - P_i = u^i (u^i)^\top$. Thus, using $\|ab^\top\|_2 = \|b\|_2$ for any rank-one matrix $ab^\top$ with $\|a\|_2 = 1$, we obtain

$$\|(I - P_1)P_2\|_2 = \|u^1 (u^1 - \alpha u^2)^\top\|_2 = \sqrt{1 - \alpha^2},$$

whence

$$\|u^1 - u^2\|_2 = \sqrt{2(1-\alpha)} \leq \frac{2}{1+\alpha}\sqrt{1-\alpha^2} = \frac{2}{1+\alpha}\|(I - P_1)P_2\|_2. \tag{7.16}$$

Now

$$\begin{aligned}(I-P_1)P_2 &= (I-P_1)P_2^\top = (I-P_1)A_2^\top(A_2^+)^\top\\ &= (I-P_1)[A_1^\top + (A_2-A_1)^\top](A_2^+)^\top\\ &= (I-P_1)(A_2-A_1)^\top(A_2^+)^\top\end{aligned}$$

together with (7.16) gives the estimate

$$\|u^1-u^2\|_2 \le \frac{2}{1+\alpha}\|(I-P_1)P_2\|_2 \le \frac{2}{1+\alpha}\|A_2^+\|_2\|A_2-A_1\|_2. \tag{7.17}$$

Analogously, we obtain

$$\|u^2-u^1\|_2 \le \frac{2}{1+\alpha}\|(I-P_2)P_1\|_2 \le \frac{2}{1+\alpha}\|A_1^+\|_2\|A_1-A_2\|_2, \tag{7.18}$$

and, together, the inequalities (7.17) and (7.18) prove the result. □

Recall that a mapping $G : E \subset V \mapsto W$ from some open subset of a finite-dimensional normed linear space V into another such space W is *locally Lipschitz* on E if for any point $x^0 \in E$ there exists a closed ball $\bar{B}(x^0,\delta) \subset E$, $\delta > 0$, and a constant $c > 0$ such that

$$\|Gx - Gy\| \le c\|x-y\|, \quad \forall\, x,y \in B(x^0,\delta).$$

A C^1 map on E is necessarily locally Lipschitz as follows readily by an application of the integral mean-value theorem. Evidently, if G is locally Lipschitz then it is Lipschitz continuous on any compact subset of E.

The principal result of this section can now be phrased as follows.

THEOREM 7.6. *Suppose that* $F : E \mapsto \mathrm{R}^{n-1}$ *is* C^1 *on some open set* $E \subset \mathrm{R}^n$ *and that* rank $DF(x) = n-1$, $\forall x \in E$. *Then, for each* $x \in E$ *there exists a unique* $u_x \in \mathrm{R}^n$ *such that*

$$DF(x)u_x = 0,\ \|u_x\|_2 = 1,\ \det\begin{pmatrix} DF(x)\\ u_x^\top\end{pmatrix} > 0, \tag{7.19}$$

and the mapping

$$G : E \mapsto \mathrm{R}^n, \quad Gx = u_x, \quad \forall x \in E, \tag{7.20}$$

is locally Lipschitz on E.

Proof. Clearly, by the rank assumption we have dim ker $DF(x) = 1$ for any $x \in E$ and, hence, a unit basis vector of the nullspace is determined up to a factor ± 1. But then the determinant condition specifies the basis vector u_x uniquely and the mapping G is certainly well defined. For the proof of the local Lipschitz property consider the matrices

$$A(x,y) = \begin{pmatrix} DF(x)\\ (Gy)^\top\end{pmatrix}, \quad \forall\, x,y \in E. \tag{7.21}$$

Then for any $x \in E$ we have $\det A(x,x) > 0$; that is, $A(x,x)$ is nonsingular. Hence, by continuity of DF, there exists, for any $x^0 \in E$, a closed ball $\mathcal{B} = \bar{B}(x^0,\delta) \subset E$, $\delta > 0$ such that $\det A(x,x^0) > 0$. By shrinking δ if needed, we may assume that DF is Lipschitz continuous on $\mathcal{B}$ with constant $c > 0$. Since $A(x,x^0) = A(x,x)[I + Gx(Gx^0 - Gx)^\top]$ and $\det\,(I + ab^\top) = 1 + a^\top b$ for any rank-one matrix $ab^\top$, we have

$$\det A(x,x^0) = Gx^\top Gx^0 \det A(x,x) \tag{7.22}$$

and thus $Gx^\top Gx^0 > 0$ for $x \in \mathcal{B}$. Therefore Lemma 7.5 provides that

$$\begin{aligned} &\|Gx - Gx^0\|_2 \\ &\quad \le 2\min\left(\|DF(x)^+\|_2, \|DF(x^0)^+\|_2\right) \|DF(x) - DF(x^0)\|_2 \\ &\quad \le 2\gamma c\|x - x^0\|_2, \quad \forall\, x \in \mathcal{B}, \end{aligned}$$

where $\gamma = \max_{x\in\mathcal{B}} \|DF(x)\|_2$. Hence G is continuous at x^0. Now

$$Gx^\top Gy = 1 - Gx^\top(Gx - Gy) \ge 1 - \|Gx - Gy\|_2, \; x, y \in \mathcal{B}$$

shows that we may shrink δ such that $Gx^\top Gy \ge -1/2$ for $x, y \in \mathcal{B}$. Therefore, another use of the lemma proves that

$$\|Gx - Gy\|_2 \le 4\gamma c\|x - y\|_2, \quad \forall\, x, y \in \mathcal{B}. \;\square$$

With the mapping G of (7.20) we define now on E the autonomous initial-value problem

$$x' = Gx, \quad x(0) = x^0 \in E. \tag{7.23}$$

By the local Lipschitz continuity of G, standard ODE theory guarantees that (7.23) has for any $x^0 \in E$ a unique C^1 solution $x : J \mapsto E$ which is defined on an open interval $J \subset \mathrm{R}^1$, $0 \in J$, that is maximal with respect to set inclusion. Moreover, if $s \in \partial J$ is finite then $x(t) \to \partial E$ or $\|x(t)\|_2 \to \infty$ as $t \to s$, $t \in J$ (see, e.g., Dieudonne [77]).

Under the assumptions of Theorem 7.6 each set

$$M_b = \{x \in E \;:\; Fx = b\}, \quad b \in \text{ rge } F, \tag{7.24}$$

is a one-dimensional submanifold of R^n. A C^1-parametric curve on M_b is any C^1 function $x : J \mapsto M_b$ on an open (nonempty) interval J such that $x'(t) \ne 0$ for all $t \in J$. Let $x : J \mapsto E$ be a C^1-parametric curve on M_b and consider any C^1-parameter transformation $\psi : J \mapsto \mathrm{R}^1$ for which $\psi'(t) > 0$ for $t \in J$ and, say, $\psi(J) = \bar{J}$. Then, clearly, $y : \bar{J} \mapsto M_b$, $y = x \circ \psi$ is again a C^1-parametric curve on M_b. The relation $y \sim x$ constitutes an equivalence relation on the set of all C^1-parametric curves on M_b, and the corresponding equivalence classes define the regular C^1 curves on M_b.

If $x : J \mapsto E$ solves (7.23) on an open interval J, $0 \in J$, then

$$Fx(t) = Fx(0) + \int_0^t DF(x(\tau))x'(\tau)d\tau = Fx^0,$$

whence x is a C^1-parametric curve on M_b for $b = Fx^0$. Conversely, if $x : J \mapsto M_b$ is a C^1-parametric curve on M_b for $b = Fx^0$ then

$$DF(x(t))x'(t) = 0, \ \|x'(t)\|_2 \neq 0, \quad \forall\, t \in J. \tag{7.25}$$

We can introduce the arclength $s = \psi(t)$ as a parameter under which $\|x'(s)\|_2 = 1$ for all $s \in \psi(J)$. Hence, with s replaced by σs where $\sigma = \pm 1$ is chosen such that

$$\det \begin{pmatrix} DF(x(0)) \\ \sigma x'(0)^\top \end{pmatrix} > 0,$$

we obtain $x'(s) = Gx(s)$, $s \in \sigma\psi(J)$; that is, $x = x(s)$ solves (7.23).

Thus, using our existence result for the initial-value problem (7.23) we have proved the following result.

THEOREM 7.7. *Under the assumptions of Theorem 7.6, there exists for any $x^0 \in E$, $b = Fx^0$, a unique, regular C^1 curve on the one-dimensional manifold M_b which has no endpoint in E.*

Clearly, the regular curves on M_b can now be computed by applying a standard ODE solver to (7.23). This requires a routine for evaluating Gx for given $x \in E$. Here, the typical approach is to begin with the calculation of a nonzero vector u in the nullspace of $DF(x)$ by means of any of the approaches mentioned in section 7.1. The resulting u is then normalized to Euclidean length one and multiplied by $\sigma = \pm 1$ such that $\sigma u^\top G\hat{x} \geq 0$, where $G\hat{x}$ is the computed vector at some "earlier" point $\hat{x}$. This avoids the direct implementation of the determinant condition in (7.19). The HOMPACK package of Watson, Billups, and Morgan [275] implements such an ODE approach.

Clearly, during the numerical integration of (7.23) the condition $Fx = b \equiv Fx^0$ is not explicitly enforced and the computed points will drift away from the manifold M_b. Thus, generally, this approach is not satisfactory if the aim is to generate a curve on M_b. However, there is a class of parametrized problems (7.1) where this drift is acceptable. These are the so-called *homotopy* problems. Suppose that a zero x^* of the C^1 map $F_1 : \mathrm{R}^m \mapsto \mathrm{R}^m$, $m = n - 1$, is sought. Since, in practice, all iterative solvers only converge when started sufficiently near x^*, it is desirable to "globalize the search." For this a homotopy may be introduced which connects F_1 with a mapping F_0 for which a zero x^0 is already known. In other words, we define $F : \mathrm{R}^m \times \mathrm{R}^1 \mapsto \mathrm{R}^m$ by

$$F(x, \lambda) = \lambda F_1 x + (1 - \lambda)F_0 x, \ x \in \mathrm{R}^m, \lambda \in \mathrm{R}^1. \tag{7.26}$$

Then $F(x^0, 0) = 0$ is already solved and $F(x, 1) = F_1 x = 0$ is the target system. Thus we may apply the above ODE approach to approximate the regular curve on the set of the solutions $(x, \lambda) \in \mathrm{R}^m \times \mathrm{R}^1$ of $F(x, \lambda) = 0$ starting at $(x^0, 0)$. If this curve passes through the hyperplane $\lambda = 1$ in R^n, $n = m + 1$, then the computed point $(\bar{x}, 1)$ may be expected to be near the desired endpoint $(x^*, 1)$. Hence, a standard iterative process applied to $F_1 x = 0$ and started from $\bar{x}$ will now probably converge to x^*. Of course, a priori, it is not clear whether the solution indeed passes through that hyperplane. In fact, some other homotopy of the form (7.26) may have to be constructed to ensure this.

The literature on homotopy methods has grown extensively over the years (see, e.g., the references in Morgan [182] or Watson et al. [276]). Under the (restrictive) assumption that the solution of $F(x, \lambda) = 0$ may be parametrized in terms of λ, the differential equation (7.23) simplifies to

$$D_x F(x(\lambda), \lambda) x'(\lambda) + D_\lambda F(x(\lambda), \lambda) = 0.$$

Its use for solving the parametrized nonlinear system appears to have been introduced by Davidenko in connection with a variety of problems, including integral equations, matrix inversions, determinant evaluations, matrix eigenvalue problems, as well as nonlinear systems (see, e.g., Davidenko [59], [60] and further references given in [OR]).

7.3 Continuation with Local Parametrizations

We turn now to the solution of problems (7.1) with a scalar parameter λ in the setting of Corollary 7.2. Thus, with a slight change of notation, we assume throughout this section that $F : E \mapsto \mathrm{R}^{n-1}$ is a C^r-mapping, $r \geq 1$, on some open set $E \subset \mathrm{R}^n$ and that F is a submersion on the nonempty set $M = \{x \in E \,:\, Fx = 0\}$. Then M is a one-dimensional C^r-submanifold of R^n.

Obviously, this manifold M may have several connected components. It may be noted that connected, one-dimensional differentiable manifolds admit to a complete classification as the following classical theorem shows.

THEOREM 7.8. *Any connected one-dimensional C^r-manifold, $r \geq 1$, is diffeomorphic either to the unit circle $\mathcal{S}^1$ in R^2 or to some interval on the real line; that is, to a connected subset of R^1 which is not a point.*

A proof of this result may be found, for instance, in Milnor [170]. There the theorem is stated for C^∞-manifolds, but the proof does not use this fact.

All continuation methods of the type discussed here begin from a given point x^0 on M and then produce a sequence of points x^k, $k = 0, 1, 2, \ldots$, on or near M. In principle, the step from x^k to x^{k+1} involves the construction of a local parametrization of M and the selection of a predicted point w from which a local parametrization algorithm, such as (7.12) or (7.15), converges to the desired next point x^{k+1} on M.

For the local parametrization at x^k we require a nonzero vector $v^k \in \mathrm{R}^n$ such that (7.4) holds, which here means that

$$v^k \notin \operatorname{rge} DF(x^k)^\top. \tag{7.27}$$

It is natural to call $T_x M^\perp = N_x M$ the normal space of M at x (under the natural inner product of R^n). Thus (7.27) means that v^k should not be a normal vector of M at x^k.

Once v^k is available, the local parametrization algorithm (7.12) requires the solution of the augmented system

$$\begin{pmatrix} Fx \\ (v^k)^\top (x - x^k) - y \end{pmatrix} = 0 \tag{7.28}$$

for given local coordinates $y \in \mathbb{R}^1$. Besides the chord Newton method applied in (7.12) (and also (7.15)), other iterative processes can, of course, be used as well. In our setting the iteration starts from a point $z^k := x^k + yv^k$, $y \in \mathbb{R}^1$, but, for $k > 0$, when several points along M are already available, other starting points can also be constructed.

In summary then, three major choices are involved in the design of a continuation process, namely,

$$\text{(7.29)} \qquad \begin{array}{l} \text{(i) the coordinate direction } v^k \text{ at each step,} \\ \text{(ii) the predicted point } z^k \text{ at each step,} \\ \text{(iii) the corrector process for solving the system (7.28).} \end{array}$$

To illustrate some of the considerations entering into these choices we outline here the code PITCON of Burkardt and Rheinboldt [47], [46].

PITCON chooses the coordinate vector v^k as one of the natural basis vectors e^{i_k} of $\mathbb{R}^n$. Thus $i_k \in N_n = \{1, \dots, n\}$ represents the index of the component of x which is used as the local parameter on M near x^k. First a tangent vector u^k at x^k is computed. For this the system

$$\text{(7.30)} \qquad \begin{pmatrix} DF(x) \\ (e^i)^\top \end{pmatrix} u = e^n$$

is used with $i = i_{k-1}$ for $k > 0$ and some a priori given index i for $k = 0$ such that the matrix in (7.30) is nonsingular. Then we set

$$\text{(7.31)} \quad u^k = \frac{\sigma}{\|u\|_2} u, \ \sigma = \rho \ \text{sign}\ (u^\top e^{i_{k-1}}), \ \rho = \ \text{sign}\left(u^{k-1^\top} e^{i_{k-1}}\right), \ k > 0,$$

where ρ is given for $k = 0$. This corresponds to the definition (7.19) of the vector Gx^k with the determinant condition replaced by a comparison of the i_{k-1}th component of u and the previous tangent vector. Actually, by (7.22) we could set, for $k > 0$,

$$\text{(7.32)} \qquad \sigma = \ \text{sign}\ (u^\top e^i) \ \text{sign}\ \det \begin{pmatrix} DF(x) \\ (e^i)^T \end{pmatrix}.$$

As long as all computed points remain on a connected component of the solution manifold M this is satisfactory. But, as noted, M may well have several components. In particular, near certain bifurcation points two components may be close to each other and have opposite orientations. In such a case, the continuation process may jump from one to the other component and the use of (7.32) would force an undesirable reversal of the direction. Accordingly, PITCON always utilizes (7.31) but monitors the condition (7.32) to detect the possible presence of a (simple) bifurcation point. Of course, not all bifurcation points can be detected in this way, and any occurrence of such a signal has to be analyzed in detail to determine the geometry of that case.

Once the tangent vector u^k has been obtained, we determine the indices j_1 and j_2 of the largest and second largest component of u^k in modulus, respectively.

In general, the local parameter index $i_k \in N_n$ is set equal to j_1. But this choice may be disadvantageous if we are approaching a limit point in the j_1th variable; that is, a point $x \in M$ where $(e^{j_1})^\top Gx = 0$. Accordingly, if the three conditions

$$\text{(7.33)} \qquad \begin{cases} |(e^{j_1})^\top u^k| < |(e^{j_1})^\top u^{k-1}|, \\ |(e^{j_2})^\top u^k| > |(e^{j_2})^\top u^{k-1}|, \\ |(e^{j_2})^\top u^k| \geq \mu |(e^{j_1})^\top u^k| \end{cases}$$

are satisfied simultaneously with some fixed $\mu \in (0,1)$, then we set $i_k = j_2$. Of course, if there is no prior tangent vector, then this check is bypassed. It should be noted that neither the choice of the indices j_1 and j_2 nor the relations (7.33) are invariant under scalings of the variables. This points to the need, observed in all continuation processes, for an a priori scaling of the original problem such that the variation of all variables is of about the same order of magnitude. In PITCON a poor scaling may be exhibited by a constant parameter index i_k for many steps and convergence failures.

Now a predicted point $z^k = x^k + h_k u^k$ along the tangent direction is computed with some step size $h_k > 0$. In order to estimate the distance between the tangent line $t \in \mathrm{R}^1 \mapsto \pi(t) = x^k + t$ and the manifold M we introduce the quadratic Hermite–Birkhoff interpolating polynomial

$$q(t) = x^k + tu^k - \frac{1}{2}t^2 w^k, \; w^k = \frac{1}{\Delta s_k}(u^k - u^{k-1}), \; \Delta s_k = \|x^k - x^{k-1}\|_2,$$

for which $q(0) = x^k$, $q'(0) = u^k$, $q(-\Delta s_k) = u^{k-1}$. Let $x = x(s)$ be a parametrization of M near $x(0) \doteq x^k$ in terms of the arclength. Then

$$w^k \doteq \int_0^1 x''(-r\Delta s_k)dr = x''(-\tilde{r}\Delta s_k), \; 0 < \tilde{r} < 1,$$

shows that

$$\|w^k\|_2 = \frac{2}{\Delta s_k}\left|\sin\frac{\alpha_k}{2}\right|, \qquad \alpha_k = \arccos\,((u^k)^\top u^{k-1}),$$

approximates the curvature of M at a point between $x(0)$ and $x(-\Delta s_k)$. It is tempting to derive from q a prediction of the curvature expected during the next computation step. However, it is readily apparent that for positive t the curvature defined by q has little predictive value. At best we may use the simple linear extrapolation

$$\gamma_k = \max(\gamma_{\min}, \gamma^*), \; \gamma^* = \|w^k\|_2 + \frac{\Delta s_k}{\Delta s_k - \Delta s_{k-1}}[\|w^k\|_2 - \|w^{k-1}\|_2]$$

as a prediction of the curvature during the next continuation step. Here $\gamma_{\min} > 0$ is a given small threshold.

Evidently $\|q(t) - \pi(t)\|_2 = (t^2/2)\|w^k\|_2$ represents an estimate of the distance between the tangent line and M which, for sufficiently smooth manifolds, is asymptotically correct to order three in $\max\,(|t|, \Delta s_k)$. Hence, if this distance is to be below a tolerance $\epsilon_k > 0$, then we should choose the step length as $t = \sqrt{2\epsilon/\|w^k\|_2}$. Here we replace, of course, the denominator by the predicted curvature value γ_k. Moreover, since a simple geometric consideration suggests that it is unreasonable to expect $\epsilon_k > \Delta s_k$, we define

$$\epsilon_k = \begin{cases} \epsilon_{\min}\Delta s_k, & : \ \epsilon \leq \epsilon_{\min}\Delta s_k, \\ \Delta s_k & : \ \epsilon \geq \Delta s_k, \\ \epsilon & : \ \text{otherwise}, \end{cases}$$

with given tolerances $\epsilon > \epsilon_{\min} > 0$. With this a tentative predicted step is now

$$h_k^{(1)} = \sqrt{\frac{2\epsilon_k}{\gamma_k}}.$$

It is advisable to adjust the step so as to ensure that it is approximately equal to Δs_{k+1}. There is no need to enforce this too rigidly. It suffices to define a new tentative step by the requirement $(e^i)^\top \pi(h_k^{(2)}) = (e^i)^\top q(h_k^{(1)})$, $i = i_k$ whence

$$h_k^{(2)} = h_k^{(1)} \left[1 + \frac{h_k^{(1)}}{2\Delta s_k} \left(1 - \frac{(e^1)^\top u^{k-1}}{(e^i)^\top u^k} \right) \right].$$

This formula may involve subtractive cancellation and has to be evaluated carefully. The final value h_k of the step length is now obtained from $h_k^{(2)}$ by enforcing the bounding requirements

$$\kappa_1 \Delta s_k \leq h_k \leq \kappa_2 \Delta s_k, \quad h_{\min} \leq h_k \leq h_{\max},$$

with given $0 < h_{\min} < h_{\max}$ and $0 < \kappa_1 < 1 \leq \kappa_2$. Here κ_2 may be adjusted during the process. In particular, we set $\kappa_2 = 1$ if the step to x^k was obtained only after a failure of the correction process and a corresponding reduction of the predicted step.

With this the first two choices of (7.29) have been specified and the corrector process can be started. In PITCON either the regular Newton method or the chord Newton method may be chosen. An essential aspect for any corrector process is to provide for careful convergence monitoring such that the iteration can be aborted as soon as divergence is suspected. In PITCON nonconvergence is declared if any one of the three conditions

$$\begin{aligned} &j \geq 1 \ \wedge \ \|Fy^j\| \geq \theta\|Fy^{j-1}\|, \\ &j \geq 2 \ \wedge \ \|y^j - y^{j-1}\| \geq \theta\|y^{j-1} - y^{j-2}\|, \\ &j \geq j_{\max} \end{aligned}$$

hold for the iterates $\{y^j\}$, where $\theta = 1.05$, except for $j = 1$ when $\theta = 2$ is chosen. The maximal iteration count $j_{\max}$ depends on the method, say, $j_{\max} = 8$ or $j_{\max} = 16$ for the Newton or chord Newton method, respectively. In the case of nonconvergence, the predictor step is reduced by a given factor, e.g., 1/3, unless the resulting step is below minimal step length.

The convergence tests use given tolerances τ_{abs} and τ_{rel} together with the relative error measure $\tau(y) = \tau_{\text{abs}} + \tau_{\text{rel}} \|y\|$ and the largest floating point number μ such that $f\ell(1.0 + \mu) = 1.0$. Convergence is declared if one of the conditions

$$\begin{aligned}
&j \geq 1 \ \wedge \ \|Fy^j\| \leq \tau_{\text{abs}} \ \wedge \ \|y^j - y^{j-1}\| \leq \tau(y^j), \\
&j \geq 0 \ \wedge \ \|Fy^j\| \leq 8\mu, \\
&j \geq 1 \ \wedge \ \|Fy^j\| + \|Fy^{j-1}\| < \tau_{\text{abs}} \ \wedge \ \|y^j - y^{j-1}\| \leq 8\tau(y^j), \\
&j \geq 2 \ \wedge \ \|Fy^j\| \leq 8\tau_{\text{abs}} \ \wedge \ \|y^j - y^{j-1}\| + \|y^{j-1} - y^{j-2}\| \leq \tau(y^j),
\end{aligned}$$

holds. By generating a sequence of solution points on a given curve, the continuation process reveals the shape of the curve and opens up the study of other features of the manifold M. In particular, there is usually a demand for the computation of specific target points. Beyond that, interest centers on the determination of certain types of singular points (fold-points) of M. Here we summarize only briefly the mechanism for computing targets incorporated in PITCON.

After at least one step has been taken, the program has available an old point x^{k-1}, a new point x^k, and the old tangent vector u^{k-1}. A target point x is a point on M for which the component $x_{i^*} = x^\top e^{i^*}$ for a prescribed target index $i^* \in N_n$ has a prescribed value $x_{i^*} = \bar{x}_{i^*}$. Thus, if a target calculation is requested, it is checked whether $\bar{x}_{i^*}$ lies between $(e^{i^*})^\top x^{k-1}$ and $(e^{i^*})^\top x^k$, in which case it is assumed that a solution point $x \in M$ with $x^\top e^{i^*} = \bar{x}_{i^*}$ is nearby. Then a point on the secant line $y(t) = (1-t)x^{k-1} + tx^k$, $0 \leq t \leq 1$ between x^{k-1} and x^k is determined such that $(e^{i^*})^\top y(t) = \bar{x}_{i^*}$. Now with the augmenting equation $x^\top e^{i^*} = \bar{x}_{i^*}$ the corrector process is applied, and if it terminates successfully, the resulting point is the desired target point.

This completes our summary of the continuation process PITCON. A number of other such codes have been developed (see, e.g., the survey by Allgower and Georg [5]). Instead of entering into the details of these various implementations we indicate only some of the possible choices in (7.29) that have been used. A classical choice for (7.29)(i) and (ii) is to use as coordinate vector v^k the tangent vector u^k and, for the prediction, a point $z^k = x^k + h_k u^k$ on the tangent line. Following Keller [144] this is usually called a *pseudoarclength method.* Of course, the secant direction between the current and previous point could here be chosen as well. Step-length algorithms in various settings have been discussed, e.g., by Deuflhard [71], Den Heijer and Rheinboldt [63], Georg [105], and Bank and Mittelmann [18]. Instead of linear predictors, higher-order extrapolators have also been applied (see, e.g., Mackens [160], Lundberg and Poore [159], and Watson et al. [276]). Other differences between implementations occur in the choice of the corrector process. These include update methods and inexact Newton methods (see Watson et al. [276]) as well as certain multigrid approaches (see Bank and Chan [17]). The ALCON package of Deuflhard, Fiedler, and Kunkel [72] implements a pseudoarclength method which uses a form of the tangential corrector algorithm (7.15) based on the QR-factorization of the Jacobian. In addition to these general methods, continuation algorithms have been developed for special classes of systems. In particular, Morgan [182] provides an overview of the application of homotopy approaches to the numerical solution of systems of polynomial equations.

Another related field concerns the development of methods for computing fold points of different type. The recent literature is very extensive and we refer here only to the introductory books by Keller [145] and Seydel [242], as well as to the proceedings edited by Mittelmann and Roose [173] and Seydel et al. [243] for further references.

7.4 Simplicial Approximations of Manifolds

As before suppose that $F : \mathrm{R}^n \mapsto \mathrm{R}^m$, $n = m + d$, $d \geq 1$, is a C^r map, $r \geq 1$, and a submersion on $M := \{x \in E : Fx = 0\}$. Then M is a d-dimensional C^r-submanifold of R^n or the empty set which we exclude. Obviously, for $d \geq 2$ we can apply continuation methods to compute paths on M, but it is certainly not easy to develop a good picture of a multidimensional manifold solely from information along some paths on it. This has led in recent years to the development of methods for a more direct approximation of implicitly defined manifolds of dimension exceeding one. One approach for such an approximation is to utilize some form of "triangulation" on M. For this we introduce the following concepts.

DEFINITION 7.2. (i) *An m-dimensional simplex (or simply m-simplex) σ^m in R^n, $n \geq m \geq 0$ is the closed, convex hull, $\sigma^m = \mathrm{co}\,(u^0, \cdots, u^m)$, of $m+1$ points $u^0, \cdots, u^m \in \mathrm{R}^n$ that are in general position. These points form the* vertex set vert $(\sigma^m) = \{u^0, \cdots, u^m\}$ *of σ^m.*
(ii) *Any $x \in \sigma^m$ can be written as*

$$x = \sum_{j=0}^{m} \xi_j u^j, \quad \sum_{j=0}^{m} \xi_j = 1, \quad \xi_j \geq 0, \quad j = 0, \ldots, m, \tag{7.34}$$

with unique barycentric coordinates $\xi_0, \ldots, \xi_m$. The point with the coordinates $\xi_j = 1/(m+1)$, $j = 0, \ldots, m$ is the barycenter of σ^m.
(iii) *The diameter of σ is* diam $(\sigma) = \max\,\{\|u^j - u^i\|_2 \;:\; i, j = 0, \ldots, m\}$.
(iv) *A k-simplex $\sigma^k \in \mathrm{R}^n$ is a k-face of σ^m if* vert $(\sigma^k) \subset$ vert (σ^m). *The unique m-face is σ_m itself and the 0-faces are the vertices.*

DEFINITION 7.3. (i) *A (finite) simplicial complex of dimension m is a finite set $\mathcal{S}$ of m-simplices*[1] *in R^n with the two properties*

(a) *if $\sigma \in \mathcal{S}$ then also all its faces belong to $\mathcal{S}$,*

(b) *for $\sigma^1, \sigma^2 \in \mathcal{S}$, $\sigma^1 \cap \sigma^2$ is either empty or a common face.*

(ii) *For a simplicial complex $\mathcal{S}$, $|\mathcal{S}| = \{x \in \mathrm{R}^n; x \in \sigma$ for some $\sigma \in \mathcal{S}\}$ is the carrier set, and* vert $(\mathcal{S}) = \{x \in \mathrm{R}^n : x \in$ vert (σ) *for some $\sigma \in \mathcal{S}\}$ the vertex set.*

[1] We exclude here complexes of simplices with different dimensions, usually permitted in combinatorial topology.

With this, a "triangulation" of a subset M_0 of a d-dimensional submanifold M of R^n is defined as a simplicial complex $\mathcal{S}$ of dimension d in R^n for which vert $(\mathcal{S}) \subset M_0$ and the points of $|\mathcal{S}|$ approximate M_0.

The task of computing such triangulations is frequently encountered in computer-aided geometric design (CAGD) and related applications. But there this task differs considerably from that considered here. In fact, in CAGD the manifolds are typically parametrically defined; that is, as the image set $\{x \in \mathrm{R}^n \,:\, x = \Phi(u, v)\}$ of some known parametrization mapping $\Phi : \mathrm{R}^2 \mapsto \mathrm{R}^n$. For that case, various triangulation methods have been proposed, all of which represent, in essence, extensions of techniques developed for the triangulation of flat spaces.

The case of implicitly defined manifolds has been addressed only fairly recently. The earliest work appears to be Allgower and Schmidt [7], Allgower and Gnutzmann [6] (see also Allgower and Georg [4]). There a piecewise affine, combinatorial continuation algorithm is used to construct a simplicial complex in the ambient space R^n that encloses the implicitly given d-dimensional manifold. The barycenters of appropriate faces of the enclosing simplices are then used to define a global, piecewise-linear approximation to the manifold. However, since in general the resulting vertices do not lie on the manifold, this does not represent a simplicial approximation in the above sense.

A first method for the direct computation of a d-dimensional simplicial complex approximating an implicitly defined manifold M in a neighborhood of a given point of M was developed by Rheinboldt [213], [214]. There standardized patches of triangulations of the tangent spaces T_xM of M are projected onto the manifold by smoothly varying projections constructed by a moving-frame algorithm. The method is applicable to manifolds of any dimension $d \geq 2$ but was used principally in the case $d = 2$. Hohmann [128] discussed a modified version of this method for the case $d = 2$, which uses interpolating polynomials to predict new points. Moreover, before correcting them onto the manifold, a search is conducted to determine whether the new points may cause a potential overlap with an already existing triangle. If such an overlap is detected, the predicted points are identified with appropriate nearby existing vertices. A further extension of the original method of Rheinboldt [213] was given by Brodzik and Rheinboldt [35]. In particular, the process was globalized to allow for the computation of a simplicial complex that approximates a specified domain on the manifold. The algorithm was then extended to the case $d > 2$ by Brodzik [34]. The implementations of these algorithms were found to be very effective for a number of practical problems.

Two different methods were developed by Melville and Mackey [166] and Henderson [125]. In both cases, interest does not center on the explicit construction of a simplicial complex on the implicitly defined, two-dimensional manifold, but on tessellating it by a cell complex. Henderson's method covers the manifold by overlapping ellipsoidal cells, each of which is obtained as the projection of a suitable ellipse on some tangent space. Melville and Mackey construct a complex of nonoverlapping cells with piecewise-curved boundaries by tracing a fishscale pattern of one-dimensional paths on the manifold. Both of these methods appear to be intrinsically restricted to two-dimensional manifolds.

We describe here briefly the principal features of the PITMAN code of Brodzik and Rheinboldt described in [35]. It assumes that the C^1 map $F : \mathrm{R}^n \mapsto \mathrm{R}^m$, $n = m + 2$ is a submersion on $M := \{x \in E \ : \ Fx = 0\}$ and hence that M is a two-dimensional submanifold of R^n. The algorithm triangulates a subset $M_0 \subset M$ of the form

$$M_0 = M \cap S, \quad S = \bigcap_k \{x \in \mathrm{R}^n \ : \ (b^k)^T(x - p^k) \geq 0\}.$$

Critical features of any algorithm of this type are the data structures for representing the mesh. For dim $M = 2$ it is still possible to work with a simple tree structure consisting of the following three arrays.

xnod: Double-precision array of dimension $mxnod \times n$ where $\mathbf{xnod}(i, j)$ stores the jth component, $j = 1, \ldots, n$, of node i.

sim: Integer array of dimension $mxsim \times 3$ where $\mathbf{sim}(i, j)$ stores the index of the jth vertex, $j = 1, 2, 3$, of simplex i.

nod: Integer array of dimension $mxnod \times mxnb + 1$ where $\mathbf{nod}(i, j)$ stores the index of the jth simplex, $j = 1, \ldots, k$, currently incident with node i and $\mathbf{nod}(i, j) = 0$ for $j = k, \ldots, mxnb$. Moreover $\mathbf{nod}(i, mxnb + 1) = \mathrm{nodtyp}$ identifies the type of the node.

The operations defined on the database are as follows.

addnod[x, *nodtyp*]: Stores the coordinates of a new point x and its index and type in the next available location of **xnod**.

addsim[x_1, x_2, x_3]: Adds a new simplex to the **sim** array, and updates the **nod** array by adding the simplex's index to the rows corresponding to the vertices x_1, x_2, x_3.

equate[x_1, x_2, $\bar{x}$]: Identifies two computed points x_1, x_2 by replacing x_1 with the projected average $\bar{x}$ of x_1 and x_2, removing x_2, and then updating all three arrays of the data structure.

chkbnd[x]: Determines whether x belongs to M_0 or not.

PITMAN uses tangential coordinate systems as defined in section 7.1, and accordingly calls on the algorithm (7.14) with $Z = DF(x)$ to set up such coordinate systems, and on (7.15) to "project" a point from a tangent space onto the manifold. The triangulation of the subset $M_0 = S \cap M$ of M begins with a user-supplied starting point $x^0 \in \mathrm{R}^n$ which need not be on M. If (7.15) fails to project x^0 onto a point $x^m \in M$, a different starting point is requested.

All computed points carry in **nod** a node type identifier nodtyp. In particular, nodtyp $= -1$ identifies a so-called frontal point which is an interior point of M_0 with an as yet incomplete simplicial neighborhood. The other values of nodtyp characterize points near and just beyond the boundary of M_0. The frontal points, in their order in **xnod**, form a queue. At each step of the process the topmost point is removed from this queue and, when new points are computed, new frontal points are added at the end. The process stops when the queue is empty. The following algorithm gives an overview of the triangulation process.

(7.35)

```
PITMAN: input: { x^0, h, amin, mxnb }
        compute the tangent basis at x^0;
        compute projection x^m of x^0 onto M;
        if x^0 ∉ M_0 then return fail;
        label x^m as a frontal point;
        compute a neighborhood of simplices of x^m;
        remove x^m from the frontal queue;
        while front is nonempty
           get the next frontal point x^c;
           compute the tangent basis at x^c;
           determine number k_c of simplices incident with x_c,
              the gap angle α, and the two open edges;
           if α < amin
             then
                call equate to identify open edges;
             else
                if k_c = mxnb then return fail;
                   else
                      find number of simplices k to fill the gap;
                      insert k simplices into gap;
                endif
           endif
           remove x^c from the list of frontal points;
        endwhile
        output: points and simplices of the triangulation
```

After the starting point $x^m \in M$ is available, the six points

$$\left(\frac{h\sqrt{3}}{2}, \frac{h}{2}\right), \ (0,h), \ \left(\frac{-h\sqrt{3}}{2}, \frac{h}{2}\right), \ \left(\frac{-h\sqrt{3}}{2}, \frac{-h}{2}\right), \ (0,-h), \ \left(\frac{h\sqrt{3}}{2}, \frac{-h}{2}\right)$$

forming the vertices of a hexagonal neighborhood of equilateral triangles around the origin in R^2 are mapped onto $x^m + T_{x^m}M$. These points are projected onto M and inserted into the queue of frontal points. For the step h either a user-supplied step size or a smaller one is applied whichever guarantees a successful projection of the first of the six points onto M. This first successful value of h is retained as the constant step size throughout the remainder of the process. The projected points inherit the connectivity pattern of the original hexagonal neighborhood of equilateral triangles in R^2.

Now the while loop begins and a frontal point x^c is taken from the queue. Then, by definition, a portion of the simplicial neighborhood of x^c has already been calculated and stored. The simplices incident with x^c define a list of edges pointing from x^c to certain other nodes. Since x^c is still a frontal point, there are exactly two among these edges that are incident only with one simplex. These edges point from x^c to some other computed nodes, say, x^{beg} and x^{end}. Hence on the tangent plane $T_{x^c}M$ of x^c they correspond to lines, to be denoted by e^{beg} and e^{end}, from the origin to (local) points $U^\top x^{beg}$ and $U^\top x^{end}$, respectively, where $U = Q_2$ is the basis matrix spanning $T_{x^c}M$ used in (7.15). The angle between e^{beg} and e^{end} defines the gap angle at x^c. Care has to be taken to use a consistent ordering of the nodes around x^c to determine whether the gap is the clockwise angle from e^{beg} to e^{end}, or its complement. We refer to the original article [35] for the details of the planar, geometric algorithm that computes the suitably oriented gap angle.

Once the gap angle α has been determined, the process of closing the gap depends on the magnitude of α, the number of already existing simplices k_c incident at x^c, and a fixed minimum acceptable angle size $amin$. If the gap is small; that is, if $\alpha < amin$, then the gap is closed by using **equate** to identify the open edges e^{beg} and e^{end}. If $\alpha \geq amin$ and k_c already equals the maximum allowable value of $mxnb$, then the algorithm stops with an error return. Otherwise,

$$k \;:=\; \max \left\{ \, i \; : \; \frac{\alpha}{i} \geq amin, \;\; 1 \leq i \leq (mxnb - k_c) \, \right\}$$

and new simplices are added to close the gap at x^c and hence to complete the neighborhood of simplices around that point. If $k = 1$, then only one simplex is added, namely the one defined by the already existing points x^c, x^{beg}, and x^{end}. The indices of these points are entered into the next row of **sim** in an order such that the orientation of the new simplex agrees with that of its adjacent simplices. If $k > 1$, then $k-1$ new points are needed to define the new simplices that close the gap. In this case, the gap angle is divided into k equal sectors, and the new points in the local coordinate system are defined to be in the resulting directions and at a distance h from x^c. One at a time, in order of rotation from e^{end} into the gap, these points are constructed in the tangent plane, projected onto M by means of (7.15), and then added to the database by **addnod**. As each new point on M is obtained, **addsim** adds to the database the simplex defined by the new point and the endpoints of the open edge from which it was rotated. At the last new point $x \in M$ in the gap, a second simplex is added, which completes the neighborhood around x^c.

For some edges e pointing to a tangent point $x^c + hp$, the algorithm (7.15) may fail to project this point onto M. In this case, a simple continuation process is started along the direction p using a smaller temporary step size until the sum of these smaller steps equals the original h. The resulting last point then becomes the desired new point to be added to the database together with the simplex that it completes. Failure of this continuation process produces an error return.

CHAPTER 8

Unconstrained Minimization Methods

As we observed, the problem of solving a system of nonlinear equations may be replaced by a problem of minimizing a nonlinear functional on R^n. The literature in this area is extensive; see, e.g., the books by Fletcher [96], Gill, Murray, and Wright [109], and Dennis and Schnabel [66], as well as the bibliography by Berman [23]. We present here only a very modest introduction to some convergence theory for a few classes of methods for solving unconstrained minimization problems.

Throughout this chapter, unless otherwise specified, it will be assumed that $g : R^n \mapsto R^1$ is a given C^1 functional on R^n. This is a more stringent condition than is really needed; in fact, in many instances it would suffice to require g to be G-differentiable on some open subset of R^n. For simplicity, and without much loss of generality, the ℓ_2-norm is used throughout this chapter, although the results can also be extended to other norms by using the dual norm on $L(R^n, R^1)$ where appropriate.

8.1 Admissible Step-Length Algorithms

Many iterative methods for constructing a (local or global) minimizer of g have the form

$$x^{k+1} = x^k - \omega_k \alpha_k p^k, \qquad k = 0, 1, \ldots \tag{8.1}$$

involving a direction vector $p^k \in R^n$, a step length $\alpha_k \geq 0$, and a relaxation parameter $\omega_k \geq 0$. The process is called a *general descent method* if

$$g(x^k) \geq g(x^{k+1}), \qquad k = 0, 1, \ldots. \tag{8.2}$$

Here we may distinguish two conceptual approaches, namely, (i) to consider the selection of the direction and step length as two independent tasks or (ii) to assume that their construction is interrelated. We begin with the development of a general convergence theory for a class of methods based on the first of these two approaches. The idea for this theory has been developed by several authors; see, in particular, Elkin [90], Wolfe [282], [283], Ortega and Rheinboldt [190], and [OR]. Our presentation combines and reformulates the presentation in the latter two references.

We consider the level sets of g

$$\mathcal{L}(\gamma) = \{x \in \mathrm{R}^n \ : \ g(x) \leq \gamma\}, \quad \gamma \in \mathrm{R}^n, \tag{8.3}$$

which may, of course, be empty for certain values of γ. For any $\bar{x} \in \mathcal{L}(\gamma)$ the path-connected component of $\mathcal{L}(\gamma)$ containing $\bar{x}$ will be denoted by $\mathcal{L}^c(\gamma, \bar{x})$. During a descent process, the value of g never increases, and hence for any $k \geq 0$ all further iterates will be in the level set $\mathcal{L}(g(x^k))$. In many cases, these iterates even remain in the path-connected component $\mathcal{L}^c(g(x^k), x^k)$. If this holds for all $k \geq 0$, the iteration never leads out of the set $\mathcal{L}^c(g(x^0), x^0)$. Observe that when $\mathcal{L}^c(\gamma, \bar{x}) \neq \emptyset$ is bounded, and therefore compact, then g cannot be a linear functional; that is, $Dg(x)$ cannot be constant on R^n.

Clearly, given a current point x, we want to use a direction vector p such that for some $\delta > 0$ we have $x - tp \in \mathcal{L}(g(x))$ for $t \in [0, \delta)$. From $\lim_{t \to 0}(1/t)[g(x) - g(x - tp)] = Dg(x)p$ it follows that, in order for this to hold for $p \neq 0$, it is sufficient that $Dg(x)p > 0$ and necessary that $Dg(x)p \geq 0$. Accordingly, we call a vector $p \neq 0$ an *admissible direction* of g at a point x if $Dg(x)p > 0$.

For the formulation of a definition of admissible steps we introduce some technical concepts, the first of which is used in defining sufficient decreases and step sizes.

DEFINITION 8.1. *A function* $\sigma : [0, \infty) \mapsto [0, \infty)$ *is a* forcing function *if for any sequence* $\{t_k\} \subset [0, \infty)$, *it follows from* $\lim_{k \to \infty} \sigma(t_k) = 0$ *that* $\lim_{k \to \infty} t_k = 0$.

In many cases the required forcing functions are derived from the *reverse modulus of continuity* of Dg.

DEFINITION 8.2. *The reverse modulus of continuity of* Dg *on a nonempty set* $E \subset \mathrm{R}^n$ *is the function* $\delta : [0, \infty) \mapsto [0, \infty)$ *defined by* $\delta(t) = 0$ *when* $\rho = \sup_{x,y \in E} \|Dg(x) - Dg(y)\|_2 = 0$, *and by*

$$\delta(t) = \begin{cases} \inf \{\|x - y\|_2 : \|Dg(x) - Dg(y)\|_2 \geq t, \ x, y \in E\}, & t \in [0, \rho), \\ \lim_{s \to \rho-} \delta(s), & t \geq \rho, \end{cases}$$

otherwise.

Clearly, we have $\delta(0) = 0$ and δ is nondecreasing. If E is compact then Dg is uniformly continuous on E, and in that case, except when Dg is constant on E, we have $\delta(t) > 0$ for all $t > 0$. In fact, if $\delta(t) = 0$ for some $t > 0$, then, for any $\epsilon > 0$, there exist $x, y \in E$ such that $\|Dg(x) - Dg(y)\|_2 \geq t$ and $\|x - y\|_2 \leq \epsilon$, which contradicts the uniform continuity. This shows that on compact sets the reverse modulus of continuity is a forcing function provided Dg is not constant.

Now we can introduce the following concept of an *admissible* step length.

DEFINITION 8.3. *A step-length algorithm* $\mathcal{T}$ *for* g *is admissible on a set* $E \subset \mathrm{R}^n$ *with relaxation range* $[\underline{\omega}, \bar{\omega}] \subset (0, \infty)$ *if the following conditions hold.*
(i) *A constant* $\mu \in (0, 1)$ *and a forcing function* $\sigma : [0, \infty) \mapsto [0, \infty)$ *are specified to measure the decrease of* g *under* $\mathcal{T}$.
(ii) $\mathcal{T}$ *accepts as admissible input any point* $x \in E$, *and any admissible, nonzero direction* $p \in \mathrm{R}^n$ *for* g *at* x.

(iii) *For any admissible input, $\mathcal{T}$ produces as output a step length $\hat{\alpha} \geq 0$ such that*

$$\hat{x} = x - \omega\hat{\alpha}p \in E, \tag{8.4}$$

$$g(x) - g(\hat{x}) \geq \omega\hat{\alpha}\mu Dg(x)p, \quad \forall\, \omega \in [\underline{\omega}, \bar{\omega}], \tag{8.5}$$

$$\hat{\alpha}\|p\|_2 \geq \sigma\left(Dg(x)\frac{p}{\|p\|_2}\right). \tag{8.6}$$

Note that (8.4) requires the next point $\hat{x}$ also to be in the domain E, while (8.5) and (8.6) call for a sufficient decrease of g and an appropriately large step, respectively. Throughout the years many different step-length algorithms have been considered which turn out to be admissible in the sense of Definition 8.3. We begin with a classical approach, the so-called principle of Curry [56], which specifies the step length as the smallest nonnegative critical point of the scalar function

$$\varphi : [0, \infty) \mapsto \mathrm{R}^1, \quad \varphi(t) = g(x^k - tp^k), \forall\, t \geq 0. \tag{8.7}$$

THEOREM 8.1. *Let $\hat{x}$ be a point in the level set $\mathcal{L}(\gamma)$ of g for which $\mathcal{L}^c(\gamma, \hat{x})$ is bounded and therefore Dg is not constant. Then, for any $\underline{\omega} \in (0, 1]$, the Curry–step-length specification*

$$\hat{\alpha} = \min\,\{\alpha \geq 0;\ Dg(x - \alpha p)p = 0\} \tag{8.8}$$

constitutes an admissible step-length algorithm for g on $\mathcal{L}^c(\gamma, \hat{x})$ with relaxation range $[\underline{\omega}, 1]$.

Proof. Let $p \neq 0$ be an admissible direction at the point $x \in \mathcal{L}^c(\gamma, \hat{x})$. Then $Dg(x)p > 0$ and

$$J = \{\alpha > 0 \ :\ g(x - tp) < g(x), \forall\, t \in (0, \alpha]\}$$

is not empty. Since $\mathcal{L}^c(\gamma, \hat{x})$ is evidently compact, $\beta = \sup_{\alpha \in J} \alpha < \infty$ exists and we have $x - tp \in \mathcal{L}^c(\gamma, \hat{x})$ for $t \in [0, \beta]$ and $g(x - \beta p) = g(x)$. This guarantees the existence of $\alpha \in (0, \beta)$ with $Dg(x - \alpha p)p = 0$. Thus the step length $\hat{\alpha}$ of (8.8) is well defined and satisfies $x - \hat{\alpha}p \in \mathcal{L}^c(\gamma, \hat{x})$, and hence also $\hat{x} = x - \omega\hat{\alpha}p \in \mathcal{L}^c(\gamma, \hat{x})$ for $\omega \in [\underline{\omega}, 1]$, which is (8.4).

Since $\hat{\alpha}$ is the smallest positive root of $Dg(x - \alpha p)p = 0$, the intermediate-value theorem ensures the existence of

$$\alpha_0 = \min\left\{\alpha \in (0, \hat{\alpha}) \ :\ Dg(x - \alpha p)p = \frac{1}{2}Dg(x)p\right\}.$$

From $Dg(x - \alpha p)p > 0$ for $\alpha \in [0, \hat{\alpha})$, it follows that the scalar function (8.7) is nonincreasing for $t \in [0, \hat{\alpha}]$ and, therefore, that for some $\tilde{\alpha} \in (0, \omega\alpha_0)$,

$$\begin{aligned} g(x) - g(x - \omega\hat{\alpha}p) &\geq g(x) - g(x - \omega\alpha_0 p) \\ &= \omega\alpha_0 Dg(x - \tilde{\alpha}p)p \geq \omega\alpha_0\frac{1}{2}Dg(x)p, \quad \forall\, \omega \in [\underline{\omega}, 1]. \end{aligned} \tag{8.9}$$

This shows that (8.5) holds with $\mu = 1/2$. The definition of α_0 implies that

$$\begin{aligned}\frac{1}{2}Dg(x)\frac{p}{\|p\|_2} &= Dg(x)\frac{p}{\|p\|_2} - Dg(x-\alpha_0 p)\frac{p}{\|p\|_2} \\ &\leq \|Dg(x) - Dg(x-\alpha_0 p)\|_2.\end{aligned}$$

Hence, using the reverse modulus of continuity, we obtain

$$\hat{\alpha}\|p\|_2 \geq \alpha_0\|p\|_2 \geq \delta\left(\frac{1}{2}Dg(x)\frac{p}{\|p\|_2}\right). \tag{8.10}$$

Because Dg cannot be constant, δ is a forcing function and therefore (8.10) proves (8.6). □

The Curry principle (8.8) requires the determination of a specific solution of a one-dimensional nonlinear equation which can only be done approximately. Note, however, that the permissible relaxation range in Theorem 8.1 ensures any algorithm to be admissible that approximates the Curry step (8.8) from below with a suitably lower-bounded relative error.

The probable oldest principle for choosing a step length dates at least to Cauchy [48] and specifies the next iterate as a minimizer of the scalar function (8.7). Clearly, this particular minimizer needs to be identified more precisely; for instance, we may use the characterization

$$g(x-\hat{\alpha}p) = \min\{g(x-\alpha p)\ : \alpha \geq 0,\ x - tp \in \mathcal{L}^c(g(x), x),\ \forall\, t \in [0,\alpha]\}. \tag{8.11}$$

In this case the following admissibility result is easily deduced from the previous theorem.

THEOREM 8.2. *Let $\hat{x}$ be a point in the level set $\mathcal{L}(\gamma)$ of g for which the connected component $\mathcal{L}^c(\gamma,\hat{x})$ is bounded. Then the minimization step* (8.11) *constitutes an admissible step-length algorithm for g on $\mathcal{L}^c(\gamma,\hat{x})$ without relaxation.*

Proof. Since $\mathcal{L}^c(\gamma,\hat{x})$ is compact, a step $\hat{\alpha}$ satisfying (8.11) certainly exists. Let α_0 denote the corresponding Curry step (8.8). Then clearly $\alpha_0 \leq \hat{\alpha}$ and $g(x-\hat{\alpha}p) \leq g(x-\alpha_0 p)$ whence, by (8.9) and (8.10)

$$g(x) - g(x-\hat{\alpha}p) \geq g(x) - g(x-\alpha_0 p) \geq \frac{1}{2}\alpha_0 Dg(x)p$$

and

$$\hat{\alpha}\|p\|_2 \geq \alpha_0\|p\|_2 \geq \delta\left(Dg(x)\frac{p}{\|p\|_2}\right). \ \square$$

In general, it is here not allowed to underrelax the step length since $g(x-\alpha p) = g(x)$ for some $\alpha \in (0,\hat{\alpha})$ cannot be excluded. In order to exclude it we either need to restrict the selection of $\hat{\alpha}$ or place further assumptions upon g. An example for the latter possibility is the following convexity condition.

DEFINITION 8.4. *The functional g is quasi convex on R^n if and only if*

$$g((1-t)x + ty) \leq \max(g(x), g(y)), \quad \forall\, t \in [0,1], \quad x, y \in \mathrm{R}^n, \tag{8.12}$$

and strictly quasi convex if the strict inequality holds for any $t \in (0,1)$ whenever $x \neq y$.

It follows easily that g is quasi convex if and only if each level set $\mathcal{L}(\gamma)$ of g is convex. Hence for such functionals the level sets are path connected. It can also be shown that a continuous, strictly quasi-convex functional has at most one local minimizer which must then be global. For quasi-convex functionals we obtain the following strengthened result.

COROLLARY 8.3. *Suppose that g is quasi convex on R^n and that the level set $\mathcal{L}(\gamma)$ of g is bounded and nonempty. Then, for any $\underline{\omega} \in (0,1]$, the step-length specification*

$$g(x - \hat{\alpha}p) = \min\ \{g(x - \alpha p)\ :\ \alpha \geq 0,\ g(x - \alpha p) \leq g(x)\} \tag{8.13}$$

is admissible for g on $\mathcal{L}(\gamma)$ with relaxation range $[\underline{\omega}, 1]$.

For a strictly quasi-convex scalar function $\varphi : \mathrm{R}^1 \mapsto \mathrm{R}^1$ it is readily proved that t^* is a minimizer of φ on some interval $[a,b]$. Then, for any $a \leq s' < t' \leq b$, it follows from $\varphi(s') \geq \varphi(t')$ that $t^* \in [s', b)$ and from $\varphi(s') \leq \varphi(t')$ that $t^* \in [a, t']$. This is the basis of the so-called *golden search algorithm.* Although this is not necessary, we stay in the setting of Corollary 8.3 and assume, in addition, that g is strictly quasi convex. The following simplified version of the golden search algorithm uses as input an $x \in \mathcal{L}(\gamma)$ and a direction p which is admissible for g at x. Moreover, an initial step $\tau > 0$, a tolerance $\epsilon \in (0, 1 - \underline{\omega}]$, and some factor $q \in [1/2, 1)$ have to be supplied, and we use the scalar function φ of (8.7).

(8.14)

$\mathcal{T}$: **input:** $\{\ x, p, \tau, \epsilon, q\ \}$
while $\varphi(\tau) \geq \varphi(0)$ **do** $q := q\tau$;
$s' := 0;\ t = \tau$;
while $\varphi(t) < \varphi(s')$ **do** $s := s';\ s' := t;\ t = s' + \tau$;
while $t - s > \epsilon t$
  $s' := s + \frac{1}{2}(3 - \sqrt{5})(t - s)$;
  $t' := s + \frac{1}{2}(\sqrt{5} - 1)(t - s)$;
  if $\varphi(s') \leq \varphi(t')$ **then** $t := t'$ **else** $s := s'$;
endwhile;
return: $\hat{\alpha} := s$

The first two loops find an interval $0 < s < t$ containing the ideal step length (8.13), which we will denote here by α^*. This is possible since $p \neq 0$ is admissible and g is strictly quasi convex. Then the last loop constructs a convergent sequence of nested intervals $[s,t]$ containing α^*, and terminates with $\hat{\alpha} = s$ satisfying $\hat{\alpha} = \omega\alpha^*$ for some $\omega \in [1 - \epsilon, 1] \subset [\underline{\omega}, 1]$. Hence, it follows from Corollary 8.3 that the step-length algorithm (8.14) is admissible for g on $\mathcal{L}(\gamma)$ for the relaxation range $[\underline{\omega}, 1]$. An important feature of this algorithm was not incorporated in (8.14). In fact, when, say, $t = t'$, then the next t' equals the previous s', and the analogous result holds for $s = s'$. Therefore, after the first step, only one function evaluation is needed for each sweep through the third loop.

The step-length algorithms of Curry and Cauchy are nonconstructive in nature and, therefore, largely of theoretical interest. The golden search algorithm is constructive but restricted to a specific function class. For the practical application we should concentrate not on the exact computation of, say, a minimizer or a critical point of (8.7), but, instead, on the determination of a step length that satisfies the conditions (8.4), (8.5), and (8.6). This observation appears to occur first in the work of Armijo [10] and Goldstein [113]. It has, since then, become the basis of many modern step-length algorithms now usually identified by the name *line-search algorithms.*

In its basic form, for a given current point x and admissible direction p, the so-called Goldstein–Armijo principle requires finding a step $\alpha > 0$ such that

$$\alpha\mu_2 Dg(x)p \geq g(x) - g(x - \alpha p) \geq \alpha\mu_1 Dg(x)p, \tag{8.15}$$

where $0 < \mu_1 \leq \mu_2 < 1$ are given constants. The lower bound is to provide for an adequate decrease of g while the upper bound is intended to ensure a sufficiently large step. This principle originated as a tool for determining damping factors for Newton's method and related processes. Suppose that $\tilde{\alpha} > 0$ is some tentative step length for which the lower inequality does not hold; that is, for which $g(x) - g(x - \tilde{\alpha}p) < \tilde{\alpha}\mu_1 Dg(x)p$. Then the scalar function $\psi : [0,1] \mapsto \mathrm{R}^1$ defined by $\psi(0) = 1$ and

$$\psi(t) = \frac{g(x) - g(x - t\tilde{\alpha}p)}{t\tilde{\alpha}Dg(x)p}, \quad \forall\, t \in (0,1],$$

satisfies, by L'Hospital's rule, $\lim_{t\to 0}\psi(t) = 1$, so that ψ is continuous on $[0,1]$. Therefore, since $\psi(t) < \mu_1$, ψ takes on all values between μ_1 and 1. In particular, there exists a $t^* \in (0,1)$ so that $\mu_1 \leq \psi(t^*) \leq \mu_2$, which is (8.15) for $\alpha = t^*\tilde{\alpha}$.

This result suggests that we might start with a sufficiently large tentative step $\tilde{\alpha}$ for which the lower bound of (8.15) is violated. Then, with a given factor $q \in (0,1)$, the reduced steps $\alpha = q^j\tilde{\alpha}$ are computed recursively and the first of these values for which (8.15) holds is taken as the desired step length. Clearly, the success of this procedure depends critically on the choices of $\tilde{\alpha}$ and the constants μ_1, μ_2, and q. For instance, for closely spaced μ-values and q near zero there may not be any integer j where the process terminates; or, for large $\tilde{\alpha}$ and q near 1, we may need an inordinately large number of steps to achieve termination; or, for μ_2 close to 1, the computed step may turn out to be unacceptably small.

By itself this rudimentary algorithm does not guarantee an admissible step length. To overcome the limitations it has become standard to modify the Goldstein–Armijo principle to the following *line search principle.* For a given point x and admissible direction p, determine a step $\alpha > 0$ such that

$$g(x) - g(x - \alpha p) \geq \alpha\mu_1 Dg(x)p, \tag{8.16}$$

$$Dg(x - \alpha p)p \geq \mu_2 Dg(x)p, \tag{8.17}$$

where again $0 < \mu_1 \leq \mu_2 < 1$ are given constants. The following result shows that this indeed defines admissible step lengths.

THEOREM 8.4. *Let $\mathcal{L}(\gamma)$ be some bounded and nonempty level set of g and $0 < \mu_1 \leq \mu_2 < 1$ given constants. Consider any step algorithm $\mathcal{T}$ which accepts as input any $x \in \mathcal{L}(\gamma)$ and admissible direction $p \neq 0$ of g at x, and produces as output a step length α that satisfies* (8.16) *and* (8.17). *Then $\mathcal{T}$ is an admissible step-length algorithm on $\mathcal{L}(\gamma)$ without relaxation.*

Proof. Let $p \neq 0$ be an admissible direction at $x \in \mathcal{L}(\gamma)$ and α the corresponding output of $\mathcal{T}$. Then (8.16) implies that $x - \alpha p \in \mathcal{L}(\gamma)$ and hence that (8.4) holds. Evidently (8.16) is (8.5) with $\mu = \mu_1$. From (8.17) it follows that

$$\begin{aligned}(1-\mu_2)Dg(x)\frac{p}{\|p\|_2} &\leq Dg(x)\frac{p}{\|p\|_2} - Dg(x-\alpha p)\frac{p}{\|p\|_2}\\ &\leq \|Dg(x) - Dg(x-\alpha p)\|_2.\end{aligned}$$

Hence, as in the proof of Theorem 8.1, using the reverse modulus of continuity, we obtain

$$\alpha\|p\|_2 \geq \delta\left((1-\mu_2)Dg(x)\frac{p}{\|p\|_2}\right),$$

which, together with (8.16), proves (8.6). □

There are numerous step-length algorithms in the literature based on this principle. We follow here the methods described in Shanno and Phua [245], Fletcher [96], and Dennis and Schnabel [66]. Other variations were given by McCormick [165], and Moré and Sorensen [180]. An important aspect of these algorithms is their use of a sequence of nested intervals which enclose the desired step length and on each of which the minimizer of a suitable quadratic interpolant is computed.

For any finite interval $[\tau_0, \tau_1] \in \mathrm{R}^1$, $\tau_0 < \tau_1$, and given values q_0, q_0', q_1 let $q(t) = at^2 + bt + c$ be the unique quadratic polynomial for which $q(\tau_0) = q_0$, $q'(\tau_0) = q_0'$, and $q(\tau_1) = q_1$. If $q_1 - q_0 - q_0'(\tau_1 - \tau_0) \neq 0$, then q has a unique extremum

$$\tau_e = \tau_0 - \frac{1}{2}\frac{q_0'(\tau_1-\tau_0)^2}{q_1 - q_0 - q_0'(\tau_1-\tau_0)}.$$

The computation of this minimizer is usually safeguarded by defining $\tau^* = \tau(\tau_0, \tau_1, q_0, q_0', q_1)$ as

$$\tau^* = \begin{cases} \max\,[\tau_e, \epsilon_0] & \tau_0 = 0,\\ \min\,[\max\,[\tau_e, (1-\epsilon_1)\tau_0 + \epsilon_1\tau_1], \epsilon_1\tau_0 + (1-\epsilon_1)\tau_1] & \tau_0 > 0,\end{cases}$$

where ϵ_0, ϵ_1 are small positive constants, e.g., $\epsilon_0 = 0.1$, $\epsilon_1 = 0.2$. Moreover, for ease of notation, the value $\tau_1 = \infty$ is allowed, in which case $\tau^* = 2\tau_0$.

The following generic version of a line-search uses the setting of Theorem 8.4 and accepts as input any $x \in \mathcal{L}(\gamma)$ and admissible direction $p \neq 0$ of g at x. In addition, constants μ_1, μ_2 for the line-search principle, and safeguard constants ϵ_0, ϵ_1, are assumed to be given. We work with the scalar function φ defined by (8.7).

$\mathcal{T}$: **input:** { x, p, μ_1, μ_2, $\varphi(0) = g(x)$, $D\varphi(0) = Dg(x)p$ }
$\tau_0 := 0$; $\tau_1 := \infty$;
success := false; $\alpha = 1$;
while *success* = false
  if $\varphi(\alpha) - \varphi(0) < \mu_1 D\varphi(0)\alpha$
    then $\tau_1 := \alpha$; $\alpha := \tau(\tau_0, \tau_1, \varphi(\tau_0), D\varphi(0), \varphi(\tau_1))$;
(8.18) **else if** $D\varphi(\alpha) < \mu_2 D\varphi(0)$
    then $\tau_0 := \alpha$;
      $\alpha := \tau(\tau_0, \tau_1, \varphi(\tau_0), D\varphi(0), \varphi(\tau_1))$;
    else *success* := true;
  endif
endwhile
return: $\hat{\alpha} := \alpha$

If the initial step fails to satisfy (8.16) then it will be decreased, otherwise, if (8.16) but not (8.17) holds, then the step is increased. This is repeated until either both conditions are satisfied or we have an interval with the last two values of α as endpoints. A closer analysis shows that at the lower endpoint (8.16), but not (8.17), must hold while at the upper endpoint (8.16) fails. It then follows that an admissible step must be contained in the interior of the interval.

In the work cited on the previous page, it has been shown that the algorithm terminates after a finite number of steps. Moreover, experience indicates that, for reasonable values of the constants, relatively few steps are needed in practice. Dennis and Schnabel [66] also discuss a version of the algorithm with safeguarded cubic interpolation.

8.2 Gradient-Related Methods

Suppose that $E \subset \mathbb{R}^n$ is a compact set and that $\{x^k\} \subset E$ is any sequence for which

$$\lim_{k\to\infty} Dg(x^k)^\top = 0. \tag{8.19}$$

Then, by continuity, the limit of any convergent subsequence of $\{x^k\}$ must be a critical point of g, and hence, by the compactness of E, there exists at least one critical point of g which is an accumulation point of $\{x^k\}$. In particular, if g is known to have at most one critical point in E, then

$$\lim_{k\to\infty} x^k = x^*, \quad Dg(x^*)^\top = 0. \tag{8.20}$$

More generally, if g has only finitely many critical points in E, then it follows easily that (8.20) holds if and only if

$$\lim_{k\to\infty} (x^k - x^{k+1}) = 0. \tag{8.21}$$

Suppose now that $\mathcal{T}$ is an admissible step-length algorithm for g on E. Beginning with any $x^0 \in E$, we choose at $x^k \in E$, $k \geq 0$, an admissible direction $p^k \neq 0$ for g at x^k, and then apply $\mathcal{T}$. By Definition 8.3 we have $x^{k+1} = x^k - \omega_k \alpha_k p^k$, where for all $k \geq 0$

$$g(x^k) - g(x^{k+1}) \geq \omega_k \alpha_k \mu \|p^k\|_2 Dg(x^k) \frac{p^k}{\|p^k\|_2},$$

$$\alpha_k \|p^k\|_2 \geq \sigma \left(Dg(x^k) \frac{p^k}{\|p^k\|_2} \right),$$

and therefore, with $\sigma_0(t) = \underline{\omega} \mu t \sigma(t)$,

$$g(x^k) - g(x^{k+1}) \geq \sigma_0 \left(Dg(x^k) \frac{p^k}{\|p^k\|_2} \right) \geq 0. \tag{8.22}$$

Since E is compact, g is bounded below and it follows from (8.22) that $\lim_{k \to \infty} [g(x^k) - g(x^{k+1})] = 0$. Evidently, σ_0 is with σ also a forcing function whence

$$\lim_{k \to \infty} Dg(x^k) \frac{p^k}{\|p^k\|_2} = 0. \tag{8.23}$$

In order to conclude from (8.23) that (8.19) holds we have to restrict the choice of the directions p^k. Evidently, the angle between the direction p^k and the tangent plane of g at x^k should not tend to zero for $k \to \infty$. This is the basis of the next concept.

DEFINITION 8.5. *A nonzero direction $p \in \mathrm{R}^n$ is* gradient related *(for g) at a given point x under the forcing function $\tau : [0, \infty) \mapsto [0, \infty)$ if*

$$\left| Dg(x) \frac{p}{\|p\|_2} \right| \geq \tau(\|Dg(x)^\top\|_2).$$

Note that when x is not a critical point, then for any gradient-related direction p, either $+p$ or $-p$ is admissible for g at x. Evidently, if each step of the process uses only directions which are gradient related at x^k under a fixed forcing function, then (8.19) does follow from (8.23). This proves the following general result.

THEOREM 8.5. *Suppose that g has at most one critical point on the compact set E, and that $\mathcal{T}$ is an admissible step-length algorithm for g on E with relaxation range $[\underline{\omega}, \bar{\omega}]$. Moreover, let $\tau : [0, \infty) \mapsto [0, \infty)$ be a given forcing function. For any starting point $x^0 \in E$ construct a descent process (8.1) where at x^k the process stops when $Dg(x^k) = 0$, or else an admissible direction p^k is chosen which is gradient related under τ; the corresponding step α_k is generated by $\mathcal{T}$, and a relaxation factor $\omega_k \in [\underline{\omega}, \bar{\omega}]$ is applied. Then the resulting sequence $\{x^k\}$ remains in E and either terminates at or converges to the unique critical point of g in E.*

There are various possible generalizations of the theorem. For example, it suffices to assume that the p^k are admissible for all $k \geq 0$ but gradient related only for $k \geq k_0$ and some fixed $k_0 \geq 0$. Another extension removes the assumption that g has only one critical point in E. In fact, the result remains valid if g has finitely many critical points in E provided we can ensure that (8.21) holds. This will be discussed in the next section where it plays a more critical role.

A large class of directions is defined by a so-called quadratic model. This corresponds to the linearization approach of section 4.1 and involves approximating the functional g at the kth step by some quadratic functional

$$\left\{ \begin{array}{l} g_k(x) = g(x^k) + Dg(x^k)^\top(x - x^k) + \frac{1}{2}(x - x^k)^\top A_k(x - x^k), \\ \qquad\qquad\qquad\qquad b^k \in \mathrm{R}^n, \quad A_k \in L_S(\mathrm{R}^n), \end{array} \right. \tag{8.24}$$

which has the same value and gradient as g at x^k. Any critical point z of g_k is a solution of the linear system $A_k(z - x^k) + Dg(x^k)^\top = 0$. This suggests that we require A_k to be invertible and to use the direction $p^k = A_k^{-1}Dg(x^k)^\top$. A sufficient condition for the gradient-relatedness of this direction is given by the following result.

THEOREM 8.6. *Let $A \in L_S(\mathrm{R}^n)$ be positive definite with condition number $\kappa_2(A) = \|A\|_2\|A^{-1}\|_2$. Then for any $x \in \mathrm{R}^n$ the direction $p = A^{-1}Dg(x)^\top$ satisfies*

$$Dg(x)p \geq \gamma\|Dg(x)^\top\|_2\|p\|_2, \quad \gamma = \frac{1}{\kappa_2(A)^2} \tag{8.25}$$

and, hence, is gradient related under the linear forcing function $\tau(t) = \gamma t$.

Proof. Let $\lambda_{\max} \geq \lambda_{\min} > 0$ be the largest and smallest eigenvalues, respectively, of A. Then $h^\top Ah \geq \lambda_{\min}h^\top h$ for all $h \in \mathrm{R}^n$ and $\|A\|_2 = \lambda_{\max}$, $\|A^{-1}\|_2 = 1/\lambda_{\min}$. Therefore

$$\begin{aligned} h^\top A^{-1}h &= (A^{-1}h)^\top A(A^{-1}h) \geq \lambda_{\min}\|A^{-1}h\|_2^2 \\ &\geq \frac{\lambda_{\min}}{\lambda_{\max}^2}\|h\|_2^2 > \frac{\lambda_{\min}^2}{\lambda_{\max}^2}\|h\|_2\|A^{-1}h\|_2, \ \forall\, h \in \mathrm{R}^n, \end{aligned}$$

which, with $h = Dg(x)^\top$, gives (8.25). □

Thus, by using these directions in Theorem 8.5, we obtain the following convergence result.

THEOREM 8.7. *Suppose that g has at most one critical point on the compact set E and that $\mathcal{T}$ is an admissible step-length algorithm for g on E with relaxation range $[\underline{\omega}, \bar{\omega}]$. For any starting point $x^0 \in E$ construct a descent process* (8.1) *where at x^k the process stops when $Dg(x^k) = 0$, or else* (i) *a positive-definite matrix $A_k \in L_S(\mathrm{R}^n)$ is chosen to define the direction $p^k = A_k^{-1}Dg(x^k)^\top$,* (ii) *the corresponding step α_k is generated by $\mathcal{T}$, and* (iii) *a relaxation factor $\omega_k \in [\underline{\omega}, \bar{\omega}]$ is applied. Then the resulting iterates remain in E. Moreover, if the constructed matrices satisfy $\kappa_2(A_k) \leq \kappa < \infty$, $\forall\, k \geq 0$, then the sequence $\{x^k\}$ either terminates at or converges to the unique critical point of g in E.*

The formulation of the theorem already suggests that we might choose for the A_k the matrices B_k generated by a symmetric rank-two update process such as the DFP method (5.31) or the BFGS method (5.36). In fact, as noted in section 5.3, both of these methods retain positive-definiteness under appropriate conditions. Of course, this leaves open the question whether the condition numbers $\kappa_2(B_k)$ remain bounded for all $k \geq 0$. This is not necessarily the case and requires, in general, additional assumptions about g and the starting points. The details involve a deeper analysis of these update methods and their convergence and cannot be addressed here. We refer to the literature cited in section 5.3.

For another choice of the matrices A_k suppose that g is C^2 on R^n. Then we obtain a simple approximating quadratic (8.24) by terminating the Taylor expansion of g after the quadratic term. This means that A_k is the Hessian matrix $H_g(x) = (\partial_i \partial_j g(x),\ i,j = 1, \ldots, n)$ of g. As noted after Theorem 2.1, $H_g(x^*)$ will be positive semidefinite at any local minimizer x^* of g. In analogy to the simple zeros of nonlinear mappings (see Definition 4.1), it is reasonable to consider local minimizers of g where the Hessian matrix is positive definite. In some neighborhood $\mathcal{U}$ of such a point x^* Theorem 8.6 then gurantees that the Newton-directions $H_g(x)^{-1}Dg(x)^\top$ are gradient related under a linear forcing function. Moreover, the condition numbers $\kappa_2(H_g(x))$ remain bounded on any compact subset of $\mathcal{U}$. Thus Theorem 8.7 can be applied here with a suitable step-length algorithm. A typical result based on this setting is as follows.

THEOREM 8.8. *Suppose that g is a convex C^2 functional on the open convex set $E \subset \mathrm{R}^n$ and that the Hessian matrix $H_g(x)$ is invertible at each $x \in E$. Moreover, assume that for some $x^0 \in E$ the level set $\mathcal{L}(g(x^0))$ is bounded. Then the sequence $\{x^k\}$ generated by the damped Newton method*

$$x^{k+1} = x^k - \alpha_k H_g(x^k)^{-1} Dg(x^k)^\top, \quad k = 0, 1, \ldots \tag{8.26}$$

with the α_k constructed by any one of the step-length algorithms of section 8.1, either terminates at or converges to the unique critical point x^* of g in E which is also the unique global minimizer of g on that set.

We sketch only the simple proof which requires some elementary details about convex functionals (see, e.g., [OR]). The convexity implies that at any $x \in E$ the matrix $H_g(x)$ is positive semidefinite and hence even positive definite because of its invertibility. Moreover, all level sets are convex. Hence, the condition numbers $\kappa_2(H_g(x))$ are bounded by some constant κ on the bounded, convex set $\mathcal{L}(g(x^0))$, whence the Newton directions are admissible and gradient related under the forcing function $\tau(t) = \kappa^{-2}t$. The result now follows directly from Theorem 8.7 together with the results of section 8.1. It might be noted that under the stated assumptions the Newton process for $Dg(x)^\top = 0$ is locally convergent at x^*. From this it follows that for all sufficiently large k the descent condition will hold for $\alpha_k = 1$, in which case the asymptotic rate of convergence of (8.26) becomes that of Newton's method itself.

Theorem 8.8 admits numerous variations. We mention here only the possibility of generalizing it to the case when, in analogy to the methods of Newton form of section 4.2, $H_g(x)$ is replaced by some matrix function. In fact if $A : E \mapsto L_S(\mathrm{R}^n)$ is a continuous mapping on some compact set $E \subset \mathrm{R}^n$

and $A(x)$ is positive definite for any $x \in E$ then $\kappa_2(A(x))$ remains bounded on E and hence Theorem 8.7 applies on that set with the directions defined by $A(x)^{-1}Dg(x)^\top$. This has an application to nonlinear least-squares problems (see section 2.2), when g has the form $g(x) = (1/2)(Fx)^\top Fx$ involving a C^1 mapping $F : E \subset \mathrm{R}^n \mapsto \mathrm{R}^m$, $m \geq n$, on some open set E. By setting $A(x) = DF(x)^\top DF(x)$ and replacing F at x^k by its linearization $Fx^k + DF(x^k)(x - x^k)$, we obtain here a quadratic approximation (8.24) of g at x^k with $Dg(x^k)^\top = DF(x^k)^\top Fx^k$ and $A_k = A(x^k)$. Thus, if rank $(DF(x^k)) = n$, then the resulting directions are

$$p^k = [DF(x^k)^\top DF(x^k)]^{-1} DF(x^k)^\top Fx^k$$

and are called the Gauss–Newton directions. Note that, except for affine F, $A(x)$ differs from the Hessian matrix $H_g(x)$ which involves also the second derivative of F. But note that for $m = n$ and nonsingular $DF(x^k)$ we obtain $p^k = DF(x^k)^{-1}Fx^k$; that is, the standard Newton step for $Fx = 0$. Obviously, under the assumption rank $(DF(x)) = n$ for $x \in E$, the matrix $A(x)$ is positive definite. Thus the observations in the beginning of this paragraph apply and one readily obtains a convergence result for the damped Gauss–Newton method

$$x^{k+1} = x^k - \alpha_k [DF(x^k)^\top DF(x^k)]^{-1} DF(x^k)^\top Fx^k, \quad k = 0, 1, \ldots,$$

analogous to Theorem 8.8 (see section 14.4.4 of [OR]). The details are straightforward. Since in many least-squares problems $DF(x)$ may not have maximal rank, it is useful to introduce here a regularization in the sense of (4.5); that is, to use directions of the form

$$p^k = (A(x^k) + \lambda_k I_k)^{-1} Dg(x^k)^\top$$

with suitable positive λ_k. This is the so-called Levenberg–Marquardt method (see Levenberg [155] and Marquardt [163], and also section 8.4).

Various other direction algorithms based on quadratic approximations of g have been reported in the literature. In particular, discretizations of the Hessian matrices have been considered as well as the interpolation of g by quadratic functionals (see, e.g., Schmidt and Trinkaus [230]). This will not be pursued here.

Theorem 8.8 applies only to gradient mappings $Fx = Dg(x^k)^\top$. Clearly, there is considerable interest in algorithms for damping constants $\alpha_k \geq 0$ such that, for general C^1 mappings $F : \mathrm{R}^n \mapsto \mathrm{R}^n$, the damped Newton process

$$x^{k+1} = x^k - \alpha_k DF(x^k)^{-1} Fx^k, \quad k = 0, 1, \ldots$$

is globally convergent on suitably chosen large sets. A very effective approach for this was developed by Deuflhard [70] and has been shown to be applicable even for ill-conditioned problems; that is, for equations where $DF(x^k)$ becomes nearly singular at some iterates. Space limitations do not allow a discussion of these results.

We end here with some comments about the rate of convergence of the descent methods in this section. Clearly, such rate results will depend on the properties of the functional and on the behavior of the direction and step-length algorithms. The following very basic rate-of-convergence result was proved in [OR, (p. 477)].

THEOREM 8.9. *Let g be G-differentiable on an open set E and $\{x^k\} \subset E$ such that* $\lim_{k\to\infty} x^* \in E$. *Suppose that* (i) $Dg(x^*)^\top = 0$, (ii) *g has a second F-derivative at x^*,* (iii) *$H_g(x^*)$ is nonsingular, and (iv) there exist $\eta > 0$ and $k_0 \geq 0$ such that*

$$g(x^k) - g(x^{k+1}) \geq \eta \|Dg(x^k)^\top\|_2^2, \quad \forall\, k \geq k_0.$$

Then $R_1\{x^k\} < 1$.

This R-linear convergence is a very weak result. In fact, as we noted already in connection with Theorem 8.8, the convergence may well be much better when a local convergence result is available for the method under consideration. This has been shown by various authors, see, e.g., Toint [261], Sorensen [250], and Dennis and Schnabel [66], and we note especially the results in Dennis and Schnabel [65], Dennis and Walker [68], [69] which prove the Q-superlinear convergence of many of the methods involving symmetric rank-two updates.

8.3 Collectively Gradient-Related Directions

One of the conceptually simplest, and oldest classes of minimization processes are the univariate relaxation methods. They utilize a fixed set of vectors $u^1, \ldots, u^N$ spanning R^n and then choose, at each step, a particular one of these vectors or its negative as the next direction. In other words, we set $p^k = \pm u^i$, for some i, $1 \leq i \leq N$, where the sign is chosen so that p^k is admissible. A simple selection principle is the cyclic use of the vectors; that is, $i = k \bmod (N+1)$, $k = 0, 1, \ldots$. Another possibility is to specify u^i as a vector that provides the maximal local decrease of g; that is,

$$\left| Dg(x) \frac{u^i}{\|u^i\|_2} \right| = \max_{j=1,\ldots,N} \left| Dg(x) \frac{u^j}{\|u^j\|_2} \right|.$$

While this particular choice may turn out to be admissible and gradient related, the cyclic selection principle generally does not guarantee the resulting directions to be gradient related. In order to extend the theory of the previous section to this setting we generalize the definition of gradient-relatedness as follows.

DEFINITION 8.6. *A finite set $\{q^1, \ldots, q^m\} \subset \mathrm{R}^n$ of nonzero vectors is* collectively gradient related *(for g) at x under the forcing function $\tau : [0, \infty) \mapsto [0, \infty)$ if*

$$\max_{j=1,\ldots,m} \left| Dg(x) \frac{q^j}{\|q^j\|_2} \right| \geq \tau(\|Dg(x)^\top\|_2).$$

In a slightly different form, this concept was introduced by Ortega and Rheinboldt [190] under the name "essential gradient-relatedness."

We consider now descent processes (8.1) with the directions p^k chosen such that for some index sequence $\{k_i\}_0^\infty$ every set

$$\Pi_i = \{p^{k_i}, p^{k_i+1}, \ldots, p^{k_{i+1}-1}\}, \qquad i = 0, 1, \ldots \tag{8.27}$$

is collectively gradient related at x^{k_i} under the same forcing function $\tau : [0, \infty) \mapsto [0, \infty)$. Clearly, for $k_i = i$, $i = 0, 1, \ldots$, this reduces to the case considered in the previous section when p^k is gradient related at x^k for any $k \geq 0$.

The condition on the sets Π_i is already satisfied if we choose only the subsequence of vectors p^{k_i}, $i = 0, 1, \ldots$, as gradient-related directions at x^{k_i} under τ. This follows immediately from

$$\max_{k_i \leq j < k_{i+1}} \left| Dg(x^{k_i}) \frac{p^j}{\|p^j\|_2} \right| \geq \left| Dg(x^{k_i}) \frac{p^{k_i}}{\|p^{k_i}\|_2} \right| \geq \tau(\|Dg(x^{k_i})^\top\|_2).$$

As another example, consider a univariate relaxation method where the directions p^k are chosen from among a finite set $u^1, \ldots, u^N$ of distinct vectors or their negatives. Any finite set S of distinct vectors of R^n spanning the space defines a norm $\|x\|_S = \max\,\{|x^\top q|/\|q\|_2 \;:\; q \in S\}$ on R^n. Hence, by norm-equivalence, there is a constant $\gamma(S) > 0$ such that $\|x\|_S \geq \gamma(S)\|x\|_2$, $\forall\, x \in \mathrm{R}^n$. Now, in our process, select the indices $\{k_i\}$; that is, the sets (8.27), such that $\operatorname{span} \Pi_i = \mathrm{R}^n$, for $i = 0, 1, \ldots$. Then among the directions p^j, $j = k_i, \ldots, k_{i+1} - 1$, there is a finite subset S_i of distinct vectors spanning R^n. Since there are only finitely many possible subsets S_i, we have $\gamma = \min_{i=0,1,\ldots} \gamma(S_i) > 0$ and therefore

$$\max_{k_i \leq j < k_{i+1}} \left| Dg(x^{k_i}) \frac{p^j}{\|p^j\|_2} \right| \geq \gamma \|Dg(x^{k_i})^\top\|_2, \quad i = 0, 1 \ldots, \tag{8.28}$$

which means that the vectors of Π_i are collectively gradient related at x^{k_i} with the forcing function $\tau(t) = \gamma t$.

In the proof of Theorem 8.5 gradient-relatedness was used to conclude (8.19) from (8.23). The following theorem of Ortega and Rheinboldt [190] ensures the validity of the same conclusion in the collectively gradient-related case.

THEOREM 8.10. *Let $\{x^k\}$ be any sequence contained in a compact set $E \subset \mathrm{R}^n$ for which* (8.21) *holds. It is no restriction to assume that successive points are distinct, whence the directions $p^k = (x^{k+1} - x^k)/\|x^{k+1} - x^k\|_2$, $k \geq 0$, are well defined. Suppose that there exists an index sequence*

$$\{k_i\} \subset \{0, 1, \ldots\}, \quad 0 < k_{i+1} - k_i \leq m, \quad m > 0 \text{ fixed}, \tag{8.29}$$

such that, for any $i \geq 0$, the sets (8.27) *are collectively gradient related at x^{k_i} under a fixed forcing function τ. Then* (8.23) *implies* (8.19).

Proof. Consider the modulus of continuity

$$\omega(t) = \sup\,\{\|Dg(x) - Dg(y)\|_2; x, y \in E,\ \|x - y\|_2 \leq t\}$$

of g on E. Since the case of constant $Dg(x)$ on E is trivial, we may assume that $\omega(t) > 0$ for $t > 0$. For given $\epsilon > 0$, let k_0 be such that

$$\|x^{k+1} - x^k\|_2 \le \frac{1}{m}\epsilon, \quad \left| Dg(x^k)\frac{p^k}{\|p^k\|_2} \right| \le \omega(\epsilon), \quad k \ge k_0.$$

Then $\|x^{k+j} - x^k\|_2 \le \epsilon$ and hence $\|Dg(x^{k+j}) - Dg(x^k)\|_2 \le \omega(\epsilon)$ for $j = 1, \ldots, m$ and $k \ge k_0$. This implies that

$$\left| Dg(x^k)\frac{p^{k+j}}{\|p^{k+j}\|_2} \right| \le \|Dg(x^k) - Dg(x^{k+j})\|_2 + \left| Dg(x^{k+j})\frac{p^{k+j}}{\|p^{k+j}\|_2} \right| \le 2\omega(\epsilon)$$

and therefore

$$2\omega(\epsilon) \ge \max_{j=1,\ldots,m} \left| Dg(x^k)\frac{p^{k+j}}{\|p^{k+j}\|_2} \right| \ge \tau(\|Dg(x^k)\|_2).$$

Since $\omega(\epsilon) \to 0$ as $\epsilon \to \infty$, it follows that $\lim_{k\to\infty} Dg(x^k)^\top = 0$. □

The required validity of (8.21) holds automatically for certain step-length algorithms, for others it represents an added qualification. Following is a sufficient condition which does lead to the desired result.

THEOREM 8.11. *Let E be a compact set on which g is not constant along any line segment. Then* (8.21) *holds for any sequence $\{x^k\} \subset E$ which satisfies*

$$(8.30) \quad \begin{cases} (1-t)x^k + tx^{k+1} \in E \\ g(x^k) \ge g((1-t)x^k + tx^{k+1}) \ge g(x^{k+1}) \end{cases} \quad \forall\, t \in [0,1],\ k \ge 0.$$

Proof. Suppose that $\|x^{k_i+1} - x^{k_i}\|_2 \ge \epsilon > 0$, $\forall\, i \ge 0$, for some subsequence $\{x^{k_i}\}$. In view of the compactness we may assume that the subsequences $\{x^{k_i}\}$ and $\{x^{k_{i+1}}\}$ have limits $x^* \in E$ and $x^{**} \in E$, whence $\|x^* - x^{**}\|_2 \ge \epsilon$. Since g is bounded below on E and $\{g(x^k)\}$ is monotonically decreasing, it follows that $\lim_{k\to\infty}[g(x^k) - g(x^{k+1})] = 0$ and therefore that $g(x^*) = g(x^{**})$. Now (8.30) implies that $(1-t)x^* + tx^{**} \in E$ and $g(x^*) = g((1-t)x^* + tx^{**}) = g(x^{**})$ which contradicts the assumption that g is not constant on line segments. □

In the proof of Theorem 8.1 for the Curry step (8.8), we observed that $g(x - \alpha p)$ is a nonincreasing function of α on $[0, \hat{\alpha}]$ and hence that the Curry step satisfies (8.30). From this it follows that the same is true for the minimization step (8.11), and, if g is strictly quasi convex, for the golden search algorithm (8.14).

Theorems 8.10 and 8.11 together permit the generalization of Theorem 8.5 to the collectively gradient-related case. We shall not repeat the formulation and instead give only a global convergence theorem for a cyclic univariate relaxation process.

THEOREM 8.12. *Let g be strictly quasi convex and such that $g(x) \to \infty$ whenever $\|x\|_2 \to \infty$. With a set of vectors $u^1, \ldots, u^m$ spanning R^n consider the iteration*

$$(8.31) \quad \begin{cases} x^{k+1} = x^k - \omega_k\alpha_k \mathrm{sign}\,(Dg(x^k)u^i)\, u^i, \\ \qquad\qquad i = k \bmod (m+1),\ k = 0, 1, \ldots, \end{cases}$$

where α_k *is obtained either by* (8.8), (8.13), *or* (8.14), *and we use* $\omega_k \in [\underline{\omega}, 1]$ *with fixed* $\underline{\omega} > 0$, *in the first two cases, and* $\omega_k = 1$ *in the third case. Then the iterates are well defined and converge to the unique global minimizer of* g *in* R^n.

This result contains as a corollary a global convergence theorem for exact SOR iterations. More specifically, let $Fx = Dg(x)^\top$, and consider the exact process (6.14) with $V = \text{span } e^i$, $i = k \bmod (n+1)$; that is,

(8.32)
$$
\begin{aligned}
\mathcal{G}:\ & \textbf{input:}\{k, x^k, M_k\} \\
& i = k \bmod (n+1); \\
& \text{solve } f_i(x^k + te^i) = 0 \text{ for } t; \\
& \textbf{if } \text{ solver failed } \textbf{then return:} \text{ fail}; \\
& x^{k+1} := x^k - \omega_k t e^i; \\
& \textbf{return:}\ \ \{x^{k+1}, M_{k+1}\}
\end{aligned}
$$

Since $Dg(x)e^i = f_i(x)$, it follows that this is precisely the iteration (8.31) with the direction vectors $u^i = e^i$, $i = 1, \ldots, n$, and the underrelaxed Curry step as step-length algorithm.

8.4 Trust Region Methods

So far we considered only methods where the selection of the direction was independent of that of the step length. The so-called trust region methods take the different approach of first choosing some tentative step length and then applying the quadratic model (8.24) to determine a direction and an actual step. The literature on this topic is extensive and we can give here only a very brief conceptual introduction.

For ease of notation consider, instead of (8.24), a quadratic function

$$
(8.33) \qquad y \in \mathrm{R}^n \mapsto \varphi(y) = b^\top y + \frac{1}{2} y^\top B y, \quad b \in \mathrm{R}^n,\ B \in L_S(\mathrm{R}^n),
$$

which is zero at the origin. Of course, φ is meant to be an approximation of the given functional g near the origin. But we can "trust" this approximation only in some neighborhood of the origin; that is, in some closed ball $\bar{B}(0, \Delta)$ with an estimated radius $\Delta > 0$. Hence, we should minimize φ only on this ball; that is, we should solve the problem

$$
(8.34) \qquad \min \{\varphi(y) \ : \ y \in \mathrm{R}^n,\ \|y\|_2 \le \Delta\}.
$$

The following conditions for a solution of (8.34) were given in Sorensen [251] and Moré and Sorensen [181].

THEOREM 8.13. (i) *If* y^* *solves* (8.34) *and* $\|y^*\|_2 < \Delta$ *then* $By^* + b = 0$ *and* B *is positive semidefinite.*

(ii) *If* B *is positive semidefinite then any solution* $y^* \in \mathrm{R}^n$ *of* $By^* + b = 0$ *with* $\|y^*\|_2 \le \Delta$ *solves* (8.34).

(iii) *If* y^* *solves* (8.34) *and* $\|y^*\|_2 = \Delta$, *then there exists* $\lambda \ge 0$ *such that* $(B + \lambda I)y^* + b = 0$ *and* $B + \lambda I$ *is positive semidefinite.*

(iv) *If* $B + \lambda I$ *is positive semidefinite for some* λ *then any solution* $y^* \in \mathrm{R}^n$ *of* $(B + \lambda I)y^* + b = 0$ *with* $\|y^*\|_2 = \Delta$ *solves* (8.34) *and* $\lambda \ge 0$.

Proof. (i) If y^* solves (8.34) and $\|y^*\|_2 < \Delta$, then y^* is a global minimizer of (8.33) on the open ball $\mathcal{B} = B(0, \Delta)$ and therefore also a critical point. Hence we have $D\varphi(y^*)^\top = By^* + b = 0$, and $D^2\varphi(y^*) = B$ must be positive semidefinite (see the remarks following Theorem 2.1).

(ii) Since φ is quadratic, we have

$$(8.35)\quad \begin{cases} \varphi(y) - \varphi(y^*) = \\ \qquad D\varphi(y^*)(y - y^*) + \dfrac{1}{2}(y - y^*)^\top B(y - y^*), \ \forall\, y, y^* \in \mathrm{R}^n, \end{cases}$$

whence $\varphi(y) - \varphi(y^*) \geq D\varphi(y^*)(y - y^*)$ because B is positive semidefinite. This shows that when y^* solves $By^* + b = 0$ and satisfies $\|y^*\|_2 \leq \Delta$ then it must be a solution of (8.34).

(iii) If y^* solves (8.34) and $\|y^*\|_2 = \Delta$, then the Lagrange multiplier theorem for minimizing φ subject to the equality constraints $\gamma(y) \equiv (1/2)(y^\top y - \Delta^2) = 0$ asserts that y^* must be a critical point of the functional $(y, \lambda) \mapsto \varphi(y) + \lambda\gamma(y)$; that is,

$$(8.36)\qquad D\varphi(y^*) + \lambda D\gamma(y^*) = 0, \quad \|y^*\|_2 = \Delta,$$

or, equivalently, $(B + \lambda I)y^* + b = 0$. By assumption we have $\varphi(y) \geq \varphi(y^*)$ for all $y \in \mathcal{B}$ and thus necessarily $D\varphi(y^*)(y - y^*) \geq 0$. By (8.36) this is equivalent with $\lambda D\gamma(y^*)(y - y^*) \leq 0$ for $y \in \mathcal{B}$. Thus, in particular, we see that $0 \leq \lambda D\gamma(y^*)(-y^*) = -\lambda\|y^*\|_2$, whence $\lambda \geq 0$. It remains to show that $B + \lambda I$ is positive semidefinite. For this note that from (8.35) we obtain with $b = -(B + \lambda I)y^*$ after a short calculation that

$$(8.37)\quad \begin{cases} \varphi(y) - \varphi(y^*) = \dfrac{1}{2}\Big[(y^*)^\top y^* - y^\top y\Big] \\ \qquad\qquad + \dfrac{1}{2}(y - y^*)^\top (B + \lambda I), \quad \forall\, y \in \mathrm{R}^n. \end{cases}$$

Because $\varphi(y) \geq \varphi(y^*)$ when $\|y\|_2 = \|y^*\|_2$, it follows that

$$(y - y^*)^\top (B + \lambda I)(y - y^*) \geq 0, \quad \forall\, y \in \mathrm{R}^n, \ \|y\|_2 = \Delta,$$

which implies that $z^\top (B + \lambda I)z \geq 0$ for all $z \in \mathrm{R}^n$ as required.

(iv) Suppose that $B + \lambda I$ is positive semidefinite for some $\lambda \geq 0$ and $y^* \in \mathrm{R}^n$ satisfies $(B + \lambda I)y^* + b = 0$ and $\|y^*\|_2 = \Delta$. Then it follows again from (8.35) that (8.37) holds for all $y \in \mathrm{R}^n$ and therefore that

$$\varphi(y) - \varphi(y^*) \geq \frac{1}{2}\Big[\Delta^2 - y^\top y\Big] \geq 0, \quad \forall\, y \in \mathrm{R}^n, \ \|y\|_2 \leq \Delta.$$

In other words, y^* solves (8.34). Finally, as in the proof of (iii) we see that $\lambda \geq 0$. □

For the application of the problem (8.34) to unconstrained minimization, suppose now that $g : \mathrm{R}^n \mapsto \mathrm{R}^1$ is a C^2 functional and that we use the Newton directions. Then at the kth iterate $x = x^k$ the quadratic model (8.24) can be written in the form

$$g_k(x^k + y) = g(x^k) + \varphi(y), \quad \varphi(y) = Dg(x^k)y + \frac{1}{2}y^\top H_g(x^k)y.$$

Assume now that at this step some trust region radius $\Delta > 0$ is available. Then the idea is to compute the step $y = y^k$ from $x = x^k$ to the next iterate as an (approximate) solution of (8.34). If this step produces a satisfactory reduction of g then Δ is increased in the next step, but if the step is unsatisfactory the radius has to be decreased.

This leads to the following generic step algorithm developed by Moré and Sorensen [181]. It uses parameters $0 < \mu < \eta < 1$ and $0 < \gamma_1 < 1 < \gamma_2$ which, together with the current estimate of the trust radius Δ, are here assumed to be supplied in the memory set.

(8.38)

$\mathcal{G}$: **input:** $\{k, x^k, M_k\}$
evaluate $b := Dg(x^k)^\top$ and $B := H_g(x^k)$;
$\rho := 0$;
while $\rho \leq \mu$
  compute an approximate solution y of (8.34);
  if solver failed **then return:** fail;
  $\rho := [g(x^k + y) - g(x^k)]/\varphi(y)$;
endwhile
if $\rho \leq \eta$ **then** choose $\Delta \in [\gamma_1\Delta,\ \Delta]$;
  else choose $\Delta \in [\Delta,\ \gamma_2\Delta]$;
$x^{k+1} := x^k + y$;
return: $\{x^{k+1}, M_{k+1}\}$

Moré and Sorensen [181] developed a Newton method for computing an approximate solution of (8.34). We sketch here only the idea. In line with Theorem 8.13 the desired multiplier λ can be obtained as a zero of the scalar function

$$\psi(s) = \frac{1}{\Delta} - \frac{1}{\|H(s)\|_2}, \quad H(s) = -(B + sI)^{-1}b. \tag{8.39}$$

Suppose that for the matrix $B \in L_S(\mathbb{R}^n)$ the diagonalization $B = Q\Lambda Q^\top$, $\Lambda = \mathrm{diag}\,(\lambda_1, \ldots, \lambda_n)$, $Q^\top Q = I$ is known. Then

$$\|H(s)\|_2^{\,2} = \sum_{j=1}^{n} \frac{c_j^2}{(\lambda_j + s)^2}, \quad Q^\top b = (c_1, \ldots, c_n)^\top.$$

By differentiating $(B + sI)H(s) = b$ with respect to s it follows that

$$\psi'(s) = \frac{H(s)^\top H'(s)}{\|H(s)\|_2^{\,3}}.$$

Thus, if $s \geq 0$ is such that $B + sI$ is positive definite then a step of Newton's method applied to ψ at s involves the following computation.

(8.40)

input: $\{\ s,\ B,\ b,\ \Delta\ \}$
compute the Cholesky factorization $B + sI = R^\top R$;
solve $Rh = -b$ for $h = H(s)$;
solve $R^\top p = h$;
$s := s + [1/\Delta]\ [\|h\|_2 - \Delta]\ [\|h\|_2/\|p\|_2]^2$;
return: s

It is readily seen that ψ is convex and strictly decreasing on the interval $(-\lambda_1, \infty)$ where λ_1 denotes the smallest eigenvalues of B. Hence, when started at a point in that interval where ψ is positive, the Newton process will converge monotonically downward to a zero of ψ. Moreover, since ψ is almost linear on that interval the convergence is very rapid.

But, of course, in setting up the algorithm we had to assume that $B + sI$ is positive definite. Indeed, as long as this is the case, the algorithm will work well. Nevertheless, it is very important that the iterates are properly safeguarded to retain a positive-definite $B + sI$ and to guarantee the convergence. For the details we refer to Moré and Sorensen [181].

We cannot enter here into a discussion of the convergence theory of trust region methods. For some convergence results for the above damped Newton method see Moré and Sorensen [181] and also Dennis and Schnabel [66].

Comparisons of various implementations of line search and trust region algorithms were given by Gay [104] and Schnabel, Koontz, and Weiss [231]. Both studies show that no one code appears to be consistently more efficient or robust than any other, although there may be considerable differences between their performances.

CHAPTER 9

Nonlinear Generalizations of Several Matrix Classes

In this chapter, we discuss some classes of mappings on R^n which represent nonlinear generalizations of several closely related types of matrices arising frequently in applications. Since the topic is somewhat peripheral to the main topic of this monograph, no proofs are given. As before, it is always assumed that $F : E \subset \mathrm{R}^n \mapsto \mathrm{R}^n$ denotes a given mapping on some domain E.

9.1 Basic Function Classes

In numerical linear algebra various matrices play a particular role, as, for example, the symmetric and positive-definite ones, those with a dominant diagonal, and those with no negative elements or, alternatively, whose inverse has this property. In addition, this list must include such special classes as the M-matrices defined by Ostrowski [195], the P- and S-matrices of Fiedler and Pták [93], [94], and some of their variations.

Most of these matrices are closely related and, accordingly, they possess many similar properties. They also often occur together in certain applications, as, for instance, in connection with the solution of certain elliptic boundary-value problems or of equilibrium problems for network flows and economic models. With growing interest in nonlinear versions of these and similar problems, increasing attention has been paid to the identification and study of suitable nonlinear generalizations of the various matrices. More specifically, we consider a class $\mathcal{F}$ of mappings F to be a nonlinear generalization of a class $\mathcal{A}$ of matrices if any affine mapping $x \in \mathrm{R}^n \mapsto Ax + b$ belongs to $\mathcal{F}$ exactly if the matrix A belongs to $\mathcal{A}$. In particular, we want to identify classes $\mathcal{F}$ which inherit many of the basic properties of $\mathcal{A}$.

For some of the mentioned matrices such nonlinear generalizations are already familiar. For example, by Kerner's Theorem 2.2 the gradient mappings F are well known to represent a natural extension of the *symmetric* matrices $A \in L(\mathrm{R}^n)$. Indeed, our above qualification is met since $x \in \mathrm{R}^n \mapsto Ax + b$ is a gradient map if and only if A is symmetric.

A nonlinear generalization of positive-definiteness was introduced by Kačurovskii [138], Minty [171], and Browder [37] for operators on infinite-dimensional spaces.

Definition 9.1. *The mapping F is* monotone *if*

$$(x-y)^\top(Fx-Fy) \geq 0, \quad \forall\, x,y \in E,$$

and strictly monotone *if, in addition, the strict inequality holds whenever $x \neq y$. Moreover, F is* uniformly monotone *if*

$$(x-y)^T(Fx-Fy) \geq \mu\|x-y\|_2{}^2, \quad \forall\, x,y \in E,\ \mu > 0.$$

Clearly, the class of monotone (strictly monotone) functions includes all affine mappings induced by positive-semidefinite (positive-definite) matrices. Symmetry is not required here, and in this connection we note that when F is a gradient mapping on a convex domain E, then F is monotone (strictly monotone, uniformly monotone) exactly if the corresponding functional is convex (strictly convex, uniformly convex) (see, e.g., p. 84 of [OR]). During the past decade, the literature on monotone operators has grown rapidly (see, e.g., Aubin and Ekeland [14] for some references).

For the discussion in the remainder of this chapter, we denote by $x \leq y$ the natural (componentwise) partial ordering on R^n, while $x < y$ shall stand for $x_i < y_i$ for each $i = 1,\dots,n$. The corresponding notation is used on the space of $n \times n$ matrices.

A nonlinear generalization of the nonnegative (nonpositive) matrices gives the following functions.

Definition 9.2. *The mapping F is* isotone *if $x \leq y$, $x,y \in E$, implies $Fx \leq Fy$ and* strictly isotone *if, in addition, $Fx < Fy$ whenever $x < y$. Correspondingly, F is* antitone *if from $x \leq y$, $x,y \in E$, it follows that $Fx \geq Fy$ and* strictly antitone *if, in addition, $Fx > Fy$ whenever $x < y$.*

As a final well-known example, consider the matrices with a nonnegative inverse. Since $A \in L(\mathrm{R}^n)$ has this property if and only if $Ax \geq 0$ implies that $x \geq 0$, the natural generalization is as follows.

Definition 9.3. *The mapping F is* inverse isotone *if $Fx \leq Fy$, $x,y \in E$, implies that $x \leq y$.*

These functions were introduced by Collatz [55] under the name *operators of monotone kind*, and since then have been studied by many others, (see, e.g., Schröder [234]). The present name derives from the easily proved fact that F is inverse isotone if and only if F is injective and $F^{-1} : FE \subset \mathrm{R}^n \mapsto \mathrm{R}^n$ is isotone.

We turn now to the *M-matrices*; that is, to those matrices $A = (a_{ij}) \in L(\mathrm{R}^n)$ with a nonnegative inverse which, in addition, satisfy $a_{ij} \leq 0$ for $i \neq j$, $i,j = 1,\dots,n$. For a survey of these and related matrices we refer, e.g., to Berman and Plemmons [24]. The nonnegativity condition for the off-diagonal elements is equivalent to the requirement that $(Ax)_k \leq 0$ whenever $x \geq 0$ and $x_k = 0$. Correspondingly, the following terminology was introduced by Rheinboldt [210].

Definition 9.4. *The mapping F with the components $f_1,\dots,f_n$, is* off-diagonally antitone *if it follows from $x \leq y$, $x,y \in E$, and $x_k = y_k$ that $f_k(x) \geq f_k(y)$.*

The name reflects the fact that in this case, for any fixed x, the functions $f_k(x+te^i)$, $i \neq k$, $i = 1, \ldots, n$, are antitone in t. Nikaido [187] uses instead the term "weakly antitone." With this we are now led to the following nonlinear generalization of the M-matrices, first analyzed by Rheinboldt [210].

DEFINITION 9.5. *A mapping* $F : E \subset \mathrm{R}^n \mapsto \mathrm{R}^n$ *is an M-function if it is inverse isotone, and off-diagonally antitone.*

We illustrate this concept with two typical examples. Consider the boundary-value problem

$$u'' = \varphi(s, u, u'), \quad 0 < s < 1, \quad u(0) = \alpha, \quad u(1) = \beta,$$

and assume that the function φ is continuous on $S = [0,1] \times \mathrm{R}^2$ and has continuous partial derivatives with respect to its second and third argument which satisfy

$$\frac{\partial}{\partial u}\varphi(s,u,p) \geq 0, \qquad \left|\frac{\partial}{\partial p}\varphi(s,u,p)\right| \leq \gamma, \quad \forall\ (s,u,p)^\top \in S.$$

With $s_j = jh$, $j = 0, 1, \ldots, n+1$, $h = 1/(n+1) < 2/\gamma$, we introduce the simple discrete analogue $Fx = 0$ where $F : \mathrm{R}^n \mapsto \mathrm{R}^n$ has the components

$$(9.1) \qquad \left\{ \begin{array}{l} f_i(x) = -x_{i-1} + 2x_i - x_{i+1} + h^2\varphi\Big(s_i, x_i, \frac{1}{2h}(x_{i+1} - x_{i-1})\Big), \\ \qquad\qquad i = 1, \ldots, n, \quad x_0 = \alpha,\ x_{n+1} = \beta. \end{array} \right.$$

Then F is a continuous, surjective M-function on R^n, and hence a homeomorphism from R^n onto itself (see Rheinboldt [210]).

The second example is a nonlinear network flow problem studied by Birkhoff and Kellogg [28] and Porsching [203]. Let $\Omega = (N, \Lambda)$ be a finite, connected, directed graph as defined in section 2.3 and, for the sake of simplicity, assume that Ω is asymmetric; that is, $(i,j) \in \Lambda$ implies $(j,i) \notin \Lambda$. For any $(i,j) \in \Lambda$ we specify a continuous conductance function $\varphi_{ij} : \mathrm{R}^2 \mapsto \mathrm{R}^1$ which is strictly isotone in the first variable and strictly antitone in the second one. The net efflux from node $i \in N$ at the state $x \in \mathrm{R}^n$ is then defined as

$$(9.2) \qquad f_i(x) = \sum_{j \in \alpha^+(i)} \varphi_{ij}(x_i, x_j) - \sum_{j \in \alpha^-(i)} \varphi_{ij}(x_j, x_i), \ \forall\ i \in N,$$

where as usual

$$\alpha^+(i) = \{j \in N\ :\ (i,j) \in \Lambda\}, \quad \alpha^-(i) = \{j \in N\ :\ (j,i) \in \Lambda\}, \quad \forall\ i \in N,$$

are the positive and negative adjacency sets, respectively. We complete the formulation of the equilibrium problem by specifying on a given nonempty "boundary set" $\partial N \subset N$ the boundary conditions

$$(9.3) \qquad x_i = \gamma_i(f_i(x)), \quad \forall\ i \in \partial N,$$

with known continuous, antitone functions $\gamma_i : \mathrm{R}^1 \mapsto \mathrm{R}^1$, $i \in \partial N$. Then the mapping $F : \mathrm{R}^n \mapsto \mathrm{R}^n$ defined by (9.2), (9.3) is an M-function, (see again

Rheinboldt [210]). In connection with this example we also refer to related, and partly more general, mappings considered by Duffin [84] which are also M-functions.

Positive-definite matrices and M-matrices share a number of properties; for example, in both cases (i) all principal submatrices have a positive determinant and (ii) there exists some vector $u \geq 0$, $u \neq 0$ for which $Au > 0$. Any one of these common properties characterizes a larger class of matrices which may serve to unify the theory of the two original types of matrices. This was the approach used by Fiedler and Pták, [93], [94], who defined a P-matrix as any matrix with the property (i) and an S-matrix as one which satisfies (ii). Equivalently, a matrix $A \in L(\mathrm{R}^n)$ turns out to be a P-matrix if and only if for any $x \neq 0$ in R^n there exists an index $k = k(x)$ such that $x_k(Ax)_k > 0$.

Nonlinear generalizations of the P- and S-matrices and of their weaker forms, the P_0- and S_0-matrices, were introduced and analyzed by Rheinboldt [211] and Moré and Rheinboldt [174]. Later, Moré [177], [178] considered nonlinear versions of several further matrix classes and their applications. Here we restrict our discussion to the following nonlinear generalization of the P-matrices.

DEFINITION 9.6. *The map F is a* P-function *if for any $x, y \in E$, $x \neq y$, there exists an index $k = k(x, y)$ such that $(x_k - y_k)(f_k(x) - f_k(y)) > 0$.*

It is readily seen that F is a P-function if and only if F is injective and $F^{-1} : FE \subset \mathrm{R}^n \mapsto \mathrm{R}^n$ is a P-function. Clearly, a strictly monotone map F must be a P-function. The relation between M- and P-functions, however, is at present not known in such generality. The following result, proved by Moré and Rheinboldt [174], is restricted to rectangles in R^n; that is, Cartesian products of n intervals on the real line, each of which may be either open, closed, or semiopen and need not be bounded. Note that all of R^n is a rectangle.

THEOREM 9.1. *The mapping F is an M-function on the rectangle $Q \subset E$ if and only if F is an off-diagonally antitone P-function on Q.*

In the linear case, a symmetric P-matrix is necessarily positive definite. A corresponding nonlinear result was proved by Moré and Rheinboldt [174].

THEOREM 9.2. *If F is a continuous P-function as well as a gradient mapping on an open convex set, then F is strictly monotone on that set.*

We end this section with a comment about the only matrices still remaining on our list in the beginning of this section, namely those in the diagonally dominant class. For them nonlinear generalizations were introduced and studied by Moré [176], and we illustrate these results by considering a special case. Moré noted that $A \in L(\mathrm{R}^n)$ is strictly diagonally dominant if and only if $(Ax)_k = 0$ for some $x \neq 0$ implies that $|x_k| < \|x\|_\infty$. This leads to the following nonlinear generalization.

DEFINITION 9.7. *The mapping F is* strictly diagonally dominant *if for any $x, y \in E$, $x \neq y$, it follows from $f_k(x) = f_k(y)$ that $|x_k - y_k| < \|x - y\|_\infty$.*

Once again, these functions are necessarily injective, and, in analogy to the linear case, they are closely related to the P-functions.

THEOREM 9.3. *Let F be continuous and strictly diagonally dominant on the rectangle $Q \subset E$. Then there exists a diagonal matrix D with all diagonal entries* *equal to* ± 1 *such that* DF *is a P-function.*

In order to extend Definition 9.7 to the irreducibly diagonally dominant case and to some of its generalizations, graph theoretical considerations have to be introduced. We refer to Moré [176] for details.

9.2 Properties of the Function Classes

In this and the following section we discuss some of the many similarities between the behavior of the nonlinear mappings defined in section 9.1 and the matrices they generalize.

For any application, characterization theorems for the various functions are of considerable importance. The simplest such theorems, in concept, are obtained when differentiability is assumed. Then, in most cases, it can be shown that a function F is in a specific class if at all points of its domain the derivative belongs to the corresponding matrix class. The model result for this is, of course, Kerner's Theorem 2.2, which is not repeated in the following composite statement.

THEOREM 9.4. *Suppose that* F *is* F*-differentiable on its domain* E.

(i) *If* E *is convex, and* $DF(x)$ *is positive semidefinite (positive definite, nonnegative, positive, strictly diagonally dominant) for all* $x \in E$, *then* F *is monotone (strictly monotone, isotone, strictly isotone, strictly diagonally dominant) on* E.

(ii) *If* E *is a rectangle and* $DF(x)$ *a P-matrix (M-matrix) for all* $x \in E$, *then* F *is a P-function (M-function) on* E.

Most of part (i) is classical except for the strictly diagonally dominant case which is due to Moré [176] (where further generalizations are included). Part (ii) is given in Moré [177]. The following results show that the converses of these statements do not hold in such generality.

THEOREM 9.5. *Let* F *be G-differentiable on the open set* E.

(a) *If* F *is strictly monotone or a P-function on* E, *then, for any* $x \in E$, $DF(x)$ *is positive semidefinite or a* P_0*-matrix, respectively.*

(b) *If* F *is inverse isotone or an M-function, then, at any point* $x \in E$ *where* $DF(x)$ *is nonsingular, this derivative is inverse isotone or an M-matrix, respectively.*

The last part raises the question of whether the existence of a nonnegative inverse of $DF(x)$ at all points of a suitable domain implies the inverse isotonicity of F. The following partial answer was given by Moré and Rheinboldt [174]; for another partial result, see Rheinboldt [211].

THEOREM 9.6. *Let* F *be convex and G-differentiable on the open, convex set* E. *Then* F *is inverse isotone if and only if* $DF(x)$ *has a nonnegative inverse for each* $x \in E$.

In the nondifferentiable case considerably fewer general characterization results are known. For M-functions four related theorems were proved by Rheinboldt [210], none of which requires more than continuity. They represent nonlinear extensions of various aspects of a well-known result by Fan [91] that a matrix

$A \in L(\mathrm{R}^n)$ with nonpositive off-diagonal elements is an M-matrix if and only if $Au > 0$ for some $u > 0$. For results concerning certain types of P-functions, see also Sandberg and Willson [222], [223] and other work by the same authors.

For any positive-definite matrix, as well as for any P- or M-matrix, it is well known that any principal submatrix also belongs to the same class. In Definition 6.2 we already introduced the concept of a subfunction as a nonlinear equivalent of such submatrices. With this the cited matrix results generalize as follows.

THEOREM 9.7. (i) *If F is a P-function or a monotone (strictly monotone, strictly diagonally dominant) mapping, then any subfunction belongs to the same function class.*

(ii) *If F is an M-function on the rectangle Q, then any subfunction is again an M-function on Q.*

Part (i) follows directly from the definitions except for the strictly diagonally dominant case which is discussed by Moré [176]. Part (ii) was proved by Moré [177], and Rheinboldt [210] also showed that the subfunctions of any continuous surjective M-function F are again surjective.

These results have applications to certain discrete boundary-value problems. Briefly, suppose that F represents an equilibrium flow on a network; in other words, that the components $x_1, \ldots, x_n$ of $x \in \mathrm{R}^n$ are state variables at the n nodes of the set $N = \{1, 2, \ldots, n\}$ and that $f_i(x)$ is the net efflux from node $i \in N$ at state $x \in \mathrm{R}^n$. A simple *Dirichlet problem* then requires the determination of a state vector x which has specified values at the nodes of a boundary set $\partial N \subset N$ and for which the efflux from all the interior nodes is a prescribed quantity

$$f_i(x) = b_i, \ \forall\, i \notin \partial N, \quad x_i = b_i, \ \forall\, i \in \partial N. \tag{9.4}$$

This means that we have to solve a system of equations involving a subfunction of F corresponding to the set of interior nodes. Hence (9.4) will have at most one solution if F belongs to one of the function classes of Theorem 9.7. Results of the mentioned type for surjective M-functions also guarantee the existence of the solution. For details about this and some related, more general results involving "mixed" boundary conditions, see Rheinboldt [210], [211].

These remarks already indicate the importance of surjectivity theorems for the various types of functions. The problem is, of course, trivial for the corresponding matrix classes, since there injectivity is equivalent with bijectivity. For M-functions a number of such surjectivity results were proved by Rheinboldt [210], [211]. For all of them the underlying principle can be expressed in the form of the following theorem.

THEOREM 9.8. *A continuous, inverse isotone mapping F is surjective if and only if it is order coercive in the sense that*

$$\lim_{k\to\infty} \|Fx^k\| = \infty \quad \textit{whenever} \quad \lim_{k\to\infty} \|x^k\| = \infty$$

for any order-monotone sequence $\{x^k\} \subset \mathrm{R}^n$.

With this it is, for instance, easily verified that the M-function $F : \mathrm{R}^n \mapsto \mathrm{R}^n$ defined by (9.1) is indeed surjective. We refer to the cited articles for more details.

A result of Minty [171] states that a continuous, uniformly monotone mapping on a Hilbert space is a homeomorphism. Moré and Rheinboldt [174] proved the following stronger result for the finite-dimensional case.

THEOREM 9.9. *Let F be a continuous and uniform P-function on R^n in the sense that there exists a $\gamma > 0$ such that for any $x \neq y$ in R^n the condition*

$$(x_k - y_k)(f_k(x) - f_k(y)) \geq \gamma \|x - y\|_2^{\,2}$$

holds for some index k. Then F is a homeomorphism of R^n onto itself.

As a corollary, we obtain here not only the cited theorem of Minty, but also the following result of Sandberg and Willson [223].

COROLLARY 9.10. *Let $Fx = Ax + \Phi(x)$, $x \in \mathrm{R}^n$, where $A \in L(\mathrm{R}^n)$ is a P-matrix and $\Phi : \mathrm{R}^n \mapsto \mathrm{R}^n$ a continuous, diagonal, and isotone function on R^n. Then F is a homeomorphism from R^n onto itself.*

For another result of this type see Rheinboldt [211]. For a more general study of global injectivity and homeomorphism properties of some of these function classes see also Parthasarathy [199].

9.3 Convergence of Iterative Processes

We turn now to some convergence results about iterative processes applied to systems $Fx = b$ for which F belongs to one of our function classes.

Suppose that F is a C^1 gradient mapping as well as a P-function on the open convex set E, and that $x^* \in E$ is a simple zero of F. Then by Theorems 9.2, 9.5, and 2.2, $DF(x^*)$ is symmetric, positive definite, and hence by the Ostrowski–Reich theorem (see, e.g., Varga [269, p. 77]), the SOR matrix $H_\omega(x^*)$ of (6.4) has spectral radius less than one for $\omega \in (0, 2)$. Therefore, it follows from the discussion in sections 6.1 and 6.2 that Newton's method, the Newton–SOR processes, the nonlinear SOR iteration, as well as the SOR–Newton methods, are all locally convergent at x^* provided the relaxation factors are between zero and two. Where applicable, the corresponding block processes may also be used. As a typical example, consider, for instance, a function of the form

$$F : \mathrm{R}^n \mapsto \mathrm{R}^n,\ Fx = A\Psi(A^Tx + b) + c, \quad A \in L(\mathrm{R}^m, \mathrm{R}^n),\ \Psi : \mathrm{R}^m \mapsto \mathrm{R}^m, n \geq m,$$

arising in Maxwell's mesh and node equations (2.34) and (2.36). If A has maximal rank m and Ψ is a continuously differentiable, strictly isotone, diagonal mapping, then the above conditions are satisfied. In fact, since under the injective transformation $\hat{x} = A^Tx + b$ we have $(x - y)^\top(Fx - Fy) = (\hat{x} - \hat{y})^\top(\Psi\hat{x} - \Psi\hat{x})$, the strict monotonicity of F follows from the strict isotonicity of Ψ. Moreover, as Birkhoff [27] observed, Fx is the gradient of $g : \mathrm{R}^n \mapsto \mathrm{R}^1$, $g(x) = \varphi(A^Tx + b) + c$, where

$$\varphi : \mathrm{R}^m \mapsto \mathrm{R}^1, \quad \varphi(z) = \sum_{i=1}^{m} \int_0^{z_i} \psi_i(t)dt.$$

The above local convergence results apply also in part when F is an M-function of class C^1. In fact, if $x^* \in E$ is again a simple zero of F, then, by Theorem 9.5, $DF(x^*)$ is an M-matrix and hence $H_\omega(x^*)$ now has spectral radius less than one for $\omega \in (0, 1]$. Therefore, all the listed processes are again locally convergent at x^* provided at best underrelaxation is applied. The results also remain valid for the corresponding Jacobi methods or any of the appropriate block processes. Overrelaxation, on the other hand, requires the usual additional assumptions about $DF(x^*)$ (see, e.g., Varga [269] or Young [284]). We shall not go into details how some of these conditions, such as, for instance, p-cyclicity, may be phrased in terms of F itself.

Local convergence results of the same type may also be obtained for strictly diagonally dominant mappings and some of their generalizations. Instead, we turn to some global convergence theorems. The symmetric, positive-definite matrices, strictly diagonally dominant matrices, and M-matrices have in common that for them the Gauss–Seidel process converges. It is significant that an analogous global convergence result holds for each one of the corresponding nonlinear functions.

THEOREM 9.11. *Let F be continuous on R^n. Then the nonlinear, exact, relaxed Gauss–Seidel process* (6.11) *with $\omega_k \in [\underline{\omega}, 1]$, $\underline{\omega} > 0$, $k = 0, 1, \ldots$, converges for any $x^0 \in \mathrm{R}^n$ to the unique zero x^* of F if any one of the following conditions holds.*

(i) *F is a uniformly monotone, gradient mapping.*

(ii) *F is a surjective M-function.*

(iii) *F is a strictly diagonally dominant mapping and possesses a zero.*

None of these results is in its most general form. Part (i) is essentially the corollary discussed after Theorem 8.12 and was proved by Schechter [226], [227], together with extensions to free-steering methods and to block processes. Part (ii) was obtained by Rheinboldt [210], [211], and extends also to the Jacobi method and to block processes. Finally, part (iii) is due to Moré [176] and also holds for the Jacobi method and for the more general class of Ω-diagonally dominant mappings.

It is interesting to note that the convergence principles used in the proofs of these three global results are rather distinct. As discussed after Theorem 8.12, the nonlinear SOR process applied to gradient mappings can be viewed as a descent method for the underlying functional. With this part (i) follows easily. For parts (ii) and (iii) we write the process in the equivalent form $x^{k+1} = Gx^k$, $k = 0, 1, \ldots$. Then, in the M-function case, G turns out to be isotone and convergence derives, in essence, from the so-called lemma of Kantorovich [140] (see also section 12.5.3 of [OR]). On the other hand, for strictly diagonally dominant F, G is strictly nonexpansive under the ℓ_∞-norm, and a theorem of Diaz and Metcalf [76] can be applied.

While global convergence, of course, is very advantageous, it must be recalled that for differentiable F the asymptotic rate of convergence of the nonlinear SOR process equals that of the corresponding linear iteration for the limiting system $DF(x^*)y = b$. Thus underrelaxation is inadvisable and block processes usually improve convergence. Moreover, if overrelaxation is permitted locally near x^*, such as in case (i), adaptive strategies are desirable that generate at all steps as large an acceptable value of ω_k as possible (see Schechter [228]). Such an approach is particularly important for larger dimensions, n, since, as noted earlier, in many cases the spectral radius of the Gauss–Seidel matrix for the limiting system increases with n and hence the rate of convergence decreases. If $DF(x^*)$ is actually singular, the local theory fails. Porsching [204] has given examples of M-functions where such a case arises and, as expected, the convergence is sublinear and hence too slow for practical purposes.

We conclude this section with a few remarks about some related global convergence theorems for methods of Newton type. For Newton's method in Banach spaces, Baluev [16] proved a global result which in our terminology may be phrased as follows.

THEOREM 9.12. *Let F be a convex, inverse isotone C^1 mapping which has a zero x^*. Then Newton's method converges to x^* for any starting point $x^0 \in \mathrm{R}^n$.*

For a proof in the context of a broader theory of monotone convergence, we refer to Ortega and Rheinboldt [191] as well as to [OR], where other references are also given. For Newton–SOR methods a corresponding global result was obtained by Moré [175].

THEOREM 9.13. *Let F be a convex M-function of class C^1 on R^n, and assume that there is an M-matrix A such that $DF(x) \geq A$ for all x in R^n. Then for any $x^0 \in \mathrm{R}^n$, any sequence $\{m_k\}$ of positive integers, and any sequence $\{w_k\} \subset [\underline{\omega}, 1]$, $\underline{\omega} > 0$, the Newton–SOR iterates* (6.5) *are well defined and converge to the unique zero of F.*

This contains an earlier result of Greenspan and Parter [116] for the one-step Newton–Gauss–Seidel process applied to almost linear equations. In the case of a mapping satisfying the conditions of part (i) of Theorem 9.11, Newton's method need not converge globally. However, under these conditions Theorem 8.8 holds and hence global convergence is guaranteed if suitable damping factors are applied to the Newton steps.

CHAPTER 10

Outlook at Further Methods

Our presentation of iterative methods for solving systems of nonlinear equations is far from comprehensive. There are numerous processes which have not been mentioned, and, even for those that were considered, many interesting variants and theories have not been touched upon. This chapter presents a brief look at some examples of further classes of methods without going into any details. For some references to other results see also the proceedings edited by Allgower and Georg [3].

10.1 Higher-Order Methods

In any linearization method the nonlinear function $F : E \subset \mathrm{R}^n \mapsto \mathrm{R}^n$ is approximated near the kth iterate x^k by an affine mapping $L_k x = Fx^k + A_k(x - x^k)$, and then a solution of $L_k x = 0$ is used as x^{k+1}. Theoretically, there is no reason to restrict ourselves to linear approximations and, in fact, in the one-dimensional case, higher-order methods have always played an important role (see, e.g., Traub [265]).

Under suitable differentiability assumptions, Taylor's formula provides the approximation $Fx = P^k_m x + R(x - x^k)$ with

$$P^k_m x = Fx^k + \frac{1}{2} DF(x^k)(x - x^k) + \cdots + \frac{1}{m!} D^m F(x^k)(x - x^k)^{[m]},$$

where $D^m F(x^k)(d)^{[m]} = D^m F(x^k)(d, d, \ldots, d)$. Then, for any $m \geq 1$, a method may be defined by requiring the next iterate to be a solution of $P^k_m x = 0$. Obviously, for $m = 1$ this gives Newton's method. However, for general $m > 1$ this prescription will be impractical since mth-order polynomial equations in several variables are not at all easy to solve and, in addition, conditions are required for selecting an appropriate root.

Various recursive schemes have been proposed which transform the use of higher-order approximations into usable methods. The following approach dates back to Schröder [233] for the one-dimensional case and was discussed, in general, by Ehrmann [87]:

$\mathcal{G}$: **input:** $\{k, x^k, M_k\}$
$b^1 := 0;$
for $j = 2, \ldots, m$
$b^j := \frac{1}{2} D^2 F(x^k)(d^{j-1})^{[2]} + \cdots + \frac{1}{j!} D^j F(x^k)(d^{j-1})^{[j]};$
(10.1) solve $DF(x^k) d^j = -(Fx^k + b^j)$ for d^j;
if solver failed **then return:** *fail*;
endfor
$x^{k+1} := x^k + d^m;$
return: $\{x^{k+1}, M_{k+1}\}$

where in each step $DF(x^k)$ is assumed to be nonsingular. For $m = 1$ we obtain Newton's method, while for $m = 2$ the process has the explicit form

$$\text{(10.2)} \quad \begin{cases} x^{k+1} = x^k - DF(x^k)^{-1} F x^k \\ \qquad -\dfrac{1}{2} DF(x^k)^{-1} D^2 F(x^k)(DF(x^k)^{-1} F x^k)^{[2]}, \quad k = 0, 1, \ldots, \end{cases}$$

and is called Chebyshev's method, or the method of osculating parabolas. Using the techniques of section 4.2 we can show that, when F is C^{m+1} on an open neighborhood of the simple zero x^*, then x^* is a point of attraction of the process (10.2) and $O_R(J, x^*) \geq O_Q(J, x^*) \geq m + 1$.

The following method dates back to Halley in the 17th century for the one-dimensional case.

$\mathcal{G}$: **input:** $\{k, x^k, M_k\}$
$B^1 := 0;$
for $j = 2, \ldots, m$
$B_j := \frac{1}{2} D^2 F(x^k)(d^{j-1}) + \cdots + \frac{1}{j!} D^j F(x^k)(d^{j-1})^{[j-1]};$
(10.3) solve $[DF(x^k) + B_j] d^j = -Fx^k$ for d^j;
if solver failed **then return:** *fail*;
endfor
$x^{k+1} := x^k + d^m;$
return: $\{x^{k+1}, M_{k+1}\}$

Here all matrices $DF(x^k) + B_j$ are assumed to be invertible. For $m = 2$, this is the method of osculating hyperbolas

$$x^{k+1} = x^k - \left(DF(x^k) - \frac{1}{2} D^2 F(x^k)(DF(x^k)^{-1} F x^k) \right)^{-1} F x^k, \quad k = 0, 1, \ldots.$$

Many variations of the algorithms (10.1) and (10.3) have been discussed in the literature; see, e.g., Mertvecova [168], Necepurenko [184], Safiev [219], and Döring [80], [79], [78].

Clearly, for increasing dimension n the required computational work soon outweighs the advantage of the higher-order convergence. In order to see how little, if anything, may be gained, we compare Chebyshev's method (10.2) with a very simple method of Traub [265] derived by combining a Newton step with a modified Newton step; that is,

$$\begin{cases} y = x^k - DF(x^k)^{-1}Fx^k, \\ x^{k+1} = y - DF(x^k)^{-1}Fy. \end{cases} \tag{10.4}$$

It is easily verified that when F is C^2 in an open neighborhood of its simple zero x^*, then x^* is a point of attraction of Traub's method (10.4) and $O_R(J, x^*) \geq O_Q(J, x^*) \geq 3$. Thus the order of convergence of both methods is at least three. At each step of either process Fx^k and $DF(x^k)$ have to be evaluated, and the system with the matrix $DF(x^k)$ has to be solved twice with different right-hand sides. In addition, (10.2) requires the evaluation of $D^2F(x^k)$ as well as $\mathcal{O}(n^3)$ additions and multiplications, while (10.4) needs only one further evaluation of F as well as $\mathcal{O}(n)$ additions and no multiplications. From this viewpoint, there appears to be hardly any justification to favor the Chebyshev method for large n.

Clearly, comparisons of this type turn out to be even worse for methods with derivatives of order higher than two. Except in the case $n = 1$, where all derivatives require only one function evaluation, the practical value of methods involving more than the first derivative of F is therefore very questionable. Even in the scalar case, Ehrmann [87] shows that the effectiveness of the use of higher derivatives decreases rapidly and that in many situations at most the use of the second derivative is desirable.

10.2 Piecewise-Linear Methods

So far we have always approximated our nonlinear system of equations $Fx = 0$ by sequences of linear systems or one-dimensional equations. Another approach is to approximate the entire mapping F by a piecewise-linear mapping on R^n and to determine its zeros. Then it also becomes possible to allow for mappings F with discontinuities. The theoretical basis for this idea was developed in combinatorial topology. We summarize briefly some relevant results from two of the classical sources, namely the texts by Seifert and Threlfall [240] and Alexandroff and Hopf [2].

For this we recall the Definitions 7.2 and 7.3 of simplices and simplicial complexes and note again that the cited texts also allow simplicial complexes consisting of simplices of different dimensions which are excluded here. Since we consider only finite complexes $\mathcal{S}$, the carrier $|\mathcal{S}|$ will always be a compact subset of R^n and the diameter $\mathrm{diam}(\mathcal{S})$ is well defined as the largest diameter of the simplices of $\mathcal{S}$.

Let $\mathcal{S}$ and $\mathcal{T}$ be two simplicial complexes of R^n (not necessarily of the same dimension). A mapping $K : \mathcal{S} \mapsto \mathcal{T}$ is a *simplicial map* if it maps every simplex of $\mathcal{S}$ affinely onto a simplex of $\mathcal{T}$. Since an affine map of a simplex is fully defined by the images of its vertices, it follows that a simplicial map $K : \mathcal{S} \mapsto \mathcal{T}$ is fully defined by specifying the image $Kx \in \mathrm{vert}\ (\mathcal{T})$ of every $x \in \mathrm{vert}\ (\mathcal{S})$. Note that the images of different vertices of $\mathcal{S}$ need not be distinct. Clearly a simplicial map $K : \mathcal{S} \mapsto \mathcal{T}$ induces a mapping from $|\mathcal{S}|$ to $|\mathcal{T}|$ which is continuous. Two simplicial complexes $\mathcal{S}$ and $\mathcal{T}$ are isomorphic if there exists a bijective simplicial map between them. In that case $|\mathcal{S}|$ and $|\mathcal{T}|$ are homeomorphic.

A simplicial complex $\mathcal{T}$ is a subdivision of the simplicial complex $\mathcal{S}$ if $|\mathcal{S}| = |\mathcal{T}|$ and each simplex of $\mathcal{T}$ is contained in a simplex of $\mathcal{S}$. A subdivision of $\mathcal{S}$ is fully specified once subdivisions of each of its simplices are provided. For an example of a subdivision of an m-simplex σ, let vert $(\sigma) = \{u^0, \dots, u^m\}$ be its vertex set and u^* the barycenter. Then, for $j = 0, \dots, m$ the m-simplex σ^j with the vertex set (vert $(\sigma^m) \setminus u^j) \cup u^*$ is contained in σ and the set of these $m+1$ simplices forms a subdivision of σ, the so-called barycentric subdivision. Other important examples of subdivisions of σ were given by Freudenthal [102] (see also Kuhn [150]) and Whitney [279, pp. 358].

For any such simplex subdivision the following lemma of Sperner [253] holds.

LEMMA 10.1. *Let $\mathcal{T}$ be a subdivision of the n-simplex σ in* R^n *and assume that each vertex $x \in$* vert $(\mathcal{T})$ *is labeled with an integer $i \in \{0, \dots, n\}$ such that, in any simplex of $\mathcal{T}$ that has x as a vertex, no vertex of the facet opposite to x; that is, of the $(n-1)$-simplex with vertex set* vert $(\sigma) \setminus \{x\}$, *carries the same label i. Then there exists a completely labeled simplex in $\mathcal{T}$ which carries all the labels* $0, 1, \dots, n$.

This lemma can be used to prove the Brouwer fixed-point theorem for simplices (see, e.g., p. 377 of Alexandroff and Hopf [2]).

THEOREM 10.2. *Any continuous mapping $F : \sigma \mapsto \sigma$ of an n-simplex $\sigma \subset$* R^n *into itself has a fixed point in σ.*

By repeated subdivision complexes with arbitrarily small diameter can be generated. This is readily seen for the barycentric subdivision. The use of such subdivisions leads to the following result about approximations of continuous mappings by simplicial mappings (see Alexandroff and Hopf [2], p 318).

THEOREM 10.3. *Let $\mathcal{S}$ and $\mathcal{T}$ be simplicial complexes of* R^n *and* R^k, *respectively, and suppose that $\{\mathcal{S}^r\}_{r=1}^{\infty}$ is a sequence of successive subdivisions of $\mathcal{S}$ for which the diameter tends to zero when $r \to \infty$. If $F : |\mathcal{S}| \mapsto |\mathcal{T}|$ is a continuous mapping, then, for $\epsilon > 0$ and sufficiently large r, a simplicial map $K_r : \mathcal{S}^r \mapsto \mathcal{T}$ exists such that*

$$\max_{x \in |\mathcal{S}|} \|Fx - K_r x\|_2 \leq \epsilon.$$

Moreover, there is a continuous homotopy $\Phi : |\mathcal{S}| \times [0,1] \mapsto$ R^k *such that $\Phi(x, 0) = K_r x$, $\Phi(x, 1) = Fx$.*

The essential aspect of this discussion is that, in past decades, many of these concepts and results have been transformed into computational procedures. In particular, algorithms have been developed for determining completely labeled simplices in the setting of the Sperner Lemma 10.1 and for finding approximate fixed points of continuous mappings of a simplex into itself. Moreover, various procedures now exist that implement the statements of Theorem 10.3.

In this brief section it is impossible to attempt even a summary of these computational approaches. The first computational algorithms utilizing simplicial approximations appear to be due to Lemke and Howson [154] and Lemke [153] and addressed the numerical solution of linear complementarity problems. Then

Scarf [224] developed methods for computing approximate fixed points of continuous mappings that were not based on simplicial complexes but on a concept of primitive sets. However, a particular choice of nodes given by Scarf and Hansen [225, Ch 7] provides a direct correspondence to certain subdivisions of a simplex. Scarf's algorithms have been superceded by more efficient algorithms of Merrill [167], Eaves [85], and Eaves and Saigal [86]. The algorithm of the last-named authors approximates some homotopy by simplicial maps on suitable simplicial complexes and their subdivisions. An important area of research in recent years concerned subdivisions that have continuously varying diameter. The first of these so-called variable dimension algorithms is due to Kuhn [151], and since then numerous such algorithms have been proposed. In section 7.4 we already mentioned algorithms of Allgower, Schmidt, and Gnutzmann [7], [6] for approximating implicitly defined manifolds by simplicial mappings.

For some references to the extensive literature in the area we refer to Todd [260], and Allgower and Georg [4], [5].

10.3 Further Minimization Methods

The literature on unconstrained minimization methods is very rich and in chapter 8 we touched only upon a very modest part. In this section we mention a few additional methods of interest, again without claiming any completeness.

For linear systems the conjugate gradient iteration dates back to the early fifties (see, e.g., Hestenes and Stiefel [126]). In analogy to the GMRES process mentioned in section 6.4 we may characterize the conjugate gradient (CG) method for the solution of a symmetric positive-definite linear system $Ax = b$ as follows. The kth iterate x^k of CG is defined to be the minimizer of the quadratic problem

$$\min\,\{\varphi(x); x \in x^0 + \mathcal{K}_k\}, \quad \varphi(x) = \frac{1}{2}x^\top Ax - x^\top b,$$

where again $\mathcal{K}_k = \text{span}\,\{r^0, Ar^0, \ldots, A^{k-1}r^0\}$ is the kth Krylov subspace generated by A and the residual $r^0 = b - Ax^0$ at the starting point x^0. One of the versions of the CG iteration has the form

(10.5)

CG: **input:** $\{x^0, A, b\}$
 $x := x^0;\quad p := Ax^0 - b;$
 while convergence not achieved
  $\alpha := (Ax - b)^\top p/(Ap)^\top p;$
  $x := x - \alpha p;$
  $\beta := (Ax - b)^\top Ap/(Ap)^\top p;$
  $p := Ax - b - \beta p;$
 endwhile
 return: x

As in GMRES, in real arithmetic the CG process terminates at the solution after at most n steps. But, for the computation, the method is best set up as an iterative method which stops only when a convergence criterion is met. For details about the method in the linear case we refer, e.g., to Hackbusch [123], and Kelley [146].

The CG process has been generalized to nonlinear functionals $g : \mathbb{R}^n \mapsto \mathbb{R}^1$ by several authors. A principal advantage of the resulting algorithms is that the search directions are generated without the need for storing a matrix. Some of the earliest such methods emulate the definition of the direction vector p in (10.5). For example, the algorithm of Fletcher and Reeves [99] defines the direction vector p^{k+1} at x^{k+1} by

$$p^{k+1} = -\frac{Dg(x^{k+1})Dg(x^{k+1})^\top}{Dg(x^k)Dg(x^k)^\top}.$$

When applied to quadratic g and with the exact Cauchy step (8.11) the method reduces to (10.5). In general, the line search has to be done rather accurately to ensure convergence and, even then, periodic restarts may be advisable. For other, more general approaches and further details we refer to Gill, Murray, and Wright [109], Dennis and Turner [67], and Nazareth [183].

The algorithms of chapter 8 are designed for general C^1 functionals g. For more restricted classes of functionals special algorithms have been developed. This includes, in particular, the least-squares functionals of the form (2.20). For an introduction to the extensive literature see Gill, Murray, and Wright [109] and Dennis and Schnabel [66]. Areas of special interest include large residual problems (see, e.g., Salane [220]) and high-dimensional problems (see, e.g., Toint [262], and Thacker [259]). A variant of these problems concerns the case when data points (x_j, y_j) are to be fitted by a function $h(x, y)$ while both the sequences $\{x_j\}$ and $\{y_j\}$ are subject to experimental error (see Schwetlick and Tiller [238], and Boggs, Byrd, Donaldson, and Schnabel [31]).

Experiments with the descent methods of chapter 8 show that many of them begin to behave erratically whenever they enter regions where the derivative of g changes rapidly or fails to exist. In such cases direct search methods can be useful provided the space dimension is small. In essence, these methods generate sequences of estimates of the desired minimizer(s) of g on the basis of deterministic or probabilistic strategies which involve solely the comparison of values of the functional. Often heuristic considerations enter strongly into the design of these strategies, and the overall processes tend to defy a rigorous convergence analysis.

Some deterministic search techniques (see, e.g., Rosenbrock [217]) are related to the relaxation methods of section 8.3. Others evaluate g at the vertices of some geometrical configuration and then proceed by shifting or contracting this configuration. For example, n-cubes have been used by Box, Davies, and Swann [32], while the algorithms of Spendley, Hext, and Himsworth [252], Nelder and Mead [185], and Parkinson and Hutchinson [198] work with n-simplices. For some details of these and related processes, see Jacoby, Kowalik, and Pizzo [136].

Random search methods belong to the general class of Monte Carlo processes and cannot compare in efficiency with the descent methods of chapter 8 when g is smooth; but they have proved to be useful when there are discontinuities or the evaluation of g involves noise. Early results are due to Brooks [36] and involve a random selection of points from a given domain in search of a global minimizer. Such pure search techniques are inherently slow and all recent approaches attempt to overcome this by retaining more information during the search. For a bibliography on these directed random search methods we refer, e.g., to Heydt [127]. The following biased search algorithm of Matyas [164] illustrates some of the basic ideas behind these methods.

(10.6)

```
MATYAS: input: { x, α, β_s, β_f, σ_s, σ_f, ρ_s, ρ_f }
        b := 0;
        while x not acceptable
          output: x;
          dir := 0;
          while dir = 0
            select random direction p ∈ R^n, ||p||_2 = 1;
            y := x + αp + b;
            if g(y) < g(x) then
              b := β_s b + σ_s αp;   x = y;
              α := ρ_s α;   dir = 1;
            endif
            b := β_f b + σ_f αp;
            α := ρ_f α;
          endwhile
        endwhile
```

Some suggested parameter values are $\beta_s, \beta_f \in [1,2]$, $\sigma_s > 0$, $\sigma_f < 0$, $\beta_s + \sigma_s > 2 \geq \beta_f + \sigma_f \geq 0$. The selection of the random direction requires the evaluation of a vector $q \in \mathrm{R}^n$ of n independent normal deviates with mean zero and variance one, and the computation of $p = q/||q||_2$. An algorithm for the determination of q may be found, for instance, in Knuth [149, p. 104].

Some other random methods were developed by Schumer and Steiglitz [236], and Lawrence and Steiglitz [152]. Certain experimental results are given by Schrack and Borowski [232]. In all the test cases used there, the method of Schumer and Steiglitz and the Matyas algorithm (10.6) show a similar linear or nearly linear average rate of convergence. Scaling of the functional g has a strong effect and so does the dimension of the space, but Gaussian noise added to g does not affect the search. Generally speaking, these and other random search techniques perform best during the first part of the iteration. This suggests their possible use as starting algorithms for other processes.

A fundamental problem for all these methods is their convergence in probability, that is, the question when the iterates $\{x^k\}$ satisfy

$$\lim_{k\to\infty} P(\|x^k - x^*\| > \epsilon) = 0$$

for some minimizer x^* of g. Some such convergence results for different methods are given by Gurin and Rastrigin [121], and Matyas [164]. A second question concerns the design of measures for the efficiency of these processes. A basic concept here is the so-called search loss introduced by Rastrigin [207] and refined by Schumer and Steiglitz [236]. Broadly speaking, this search loss is the quotient of the value of g divided by the expected improvement in the function value for one function evaluation. In the case of the quadratic functional $g(x) = \|x\|_2$ this search loss function can be computed for various search processes. As shown in the cited articles this provides for a comparison between these methods and, say, simple gradient iterations.

The methods discussed in chapter 8 are local in nature in the sense that they determine a minimizer of the given functional g in the level set specified by the starting point. Often in practice the requirement arises for finding a global minimizer of g or for determining all minimizers. In recent years the literature on algorithms for these problems has been growing rapidly. In essence, all these algorithms combine a standard method for finding a local minimizer with a search procedure for further such points. Once a local minimizer has been found it has to be made "repellent" to the local minimization method and then a suitable new starting point for the local process has to be constructed. For instance, a local minimizer x of g can be made repellent by adding to g a steep hat-function which is nonzero only in some small neighborhood of x. In order to find a new starting point one has to "tunnel," in essence, through the rim of the level set specified by the previous starting point. Various techniques have been proposed for these tasks. We refer here only to the books by Ratchek and Rokne [208], Törn and Zilinskas [263], Horst and Tuy [132], Horst and Pardalos, [130], Horst, Pardalos, and Thoai [131], and Floudas and Pardalos [100].

Bibliography

[1] Abraham, R., Marsden, J. E., and Ratiu, T. *Manifolds, Tensor Analysis, and Applications*, 2nd ed., *Appl. Math. Sci.* 75, Springer-Verlag, Heidelberg, New York, 1988.

[2] Alexandroff, P., and Hopf, H. *Topologie.* Chelsea, New York, 1965, Orig. edition, Berlin, Germany, 1935.

[3] Allgower, E. L., and Georg, K., eds. *Computational Solution of Nonlinear Systems of Equations*, Lecture Notes in Appl. Math. 26, American Mathematical Society, Providence, RI, 1990.

[4] Allgower, E. L., and Georg, K. *Numerical Continuation Methods*, Ser. in Comput. Math. 13, Springer-Verlag, Heidelberg, New York, 1990.

[5] Allgower, E. L., and Georg, K. Continuation and path following. in Acta Numerica 1992, A. Iserles, ed., Cambridge University Press, Cambridge, England, 1993, pp. 1–64.

[6] Allgower, E. L., and Gnutzmann, S. An algorithm for piecewise linear approximation of an implicitly defined two-dimensional surface. *SIAM J. Numer. Anal.* 24 (1987), 452–469.

[7] Allgower, E. L., and Schmidt, P. H. An algorithm for piecewise linear approximation of an implicitly defined manifold. *SIAM J. Numer. Anal.* 22 (1985), 322–346.

[8] Ames, W. F. *Numerical Methods for Partial Differential Equations*, 3rd ed. Academic Press, New York, 1992.

[9] Aris, R. *The Mathematical Theory of Diffusion and Reaction in Permeable Catalysts.* Clarendon Press, Oxford, England, 1975.

[10] Armijo, L. Minimization of functions having Lipschitz–continuous first partial derivatives. *Pacific J. Math.* 16 (1966), 1–3.

[11] Arnoldi, W. E. The principle of minimized iterations in the solution of the matrix eigenvalue problem. *Quart. Appl. Math.* 9 (1951), 17–29.

[12] Ascher, U. M., Mattheij, R. M. M., and Russel, R. D. *Numerical solution of boundary value problems for ordinary differential equations*, Classics in Appl. Math. 13, SIAM, Philadelphia, PA, 1995.

[13] Aubin, J.-P. *Applied Functional Analysis.* John Wiley, New York, 1979.

[14] Aubin, J.-P., and Ekeland, I. *Applied Nonlinear Analysis.* John Wiley, New York, 1984.

[15] Axelson, O., and Barker, V. A. *Finite Element Solution of Boundary Value Problems.* Academic Press, New York, 1984.

[16] Baluev, A. On the method of Chaplygin. *Dokl. Akad. Nauk SSSR* 83 (1952), 781–784.

[17] Bank, R. E., and Chan, T. F. PLTMG: a multi-grid continuation program for parametrized nonlinear elliptic systems. *SIAM J. Sci. Stat. Comput.* 7 (1986), 540–559.

[18] BANK, R. E., AND MITTELMANN, H. D. Stepsize selection in continuation procedures and damped Newton's method. *J. Comput. Appl. Math.* 26 (1989), 67–77.

[19] BARNES, J. An algorithm for solving nonlinear equations based on the secant method. *Comput. J.* 8 (1965), 66–72.

[20] BARNSLEY, M. *Fractals Everywhere.* Academic Press, New York, 1988.

[21] BATES, D. M., AND WATTS, D. G. *Nonlinear Regression Analysis and its Applications.* John Wiley, New York, 1988.

[22] BERGER, M. S. *Nonlinearity and Functional Analysis.* Academic Press, New York, 1977.

[23] BERMAN, A. *Nonlinear Optimization Bibliography.* University of Waterloo Press, Waterloo, Ontario, Canada, 1985.

[24] BERMAN, A., AND PLEMMONS, R. J. *Nonnegative Matrices in the Mathematical Sciences.* Academic Press, New York, 1979.

[25] BERRY, M. W., HEATH, M. T., KANEKO, I., LAW, M., PLEMMONS, R. J., AND WARD, R. C. An algorithm to compute a sparse basis of the null space. *Numer. Math.* 47 (1985), 483–504.

[26] BERS, L. On mildly nonlinear partial difference equations of elliptic type. *J. Res. Nat. Bur. Stand. Sect. B* 51 (1953), 229–236.

[27] BIRKHOFF, G. D. A variational principle for nonlinear networks. *Quart. Appl. Math.* 21 (1963), 160–162.

[28] BIRKHOFF, G. D., AND KELLOGG, R. B. *Solution of equilibrium equations in thermal networks.* in Symposium on Generalized Networks. Brooklyn Polytechnic Press, Brooklyn, NY, 1966, pp. 443–452.

[29] BITTNER, L. Eine Verallgemeinerung des Sekantenverfahrens zur näherungsweisen Berechnung der Nullstellen eines nichtlinearen Gleichungssystems. *Wiss. Z. Tech. Univ. Dresden* 9 (1959/60), 325–329.

[30] BJÖRCK, A. *Least squares methods.* in Handbook of Numerical Analysis, P. G. Ciarlet and J. L. Lions, eds., North–Holland, Amsterdam, New York, 1990, pp. 465–647.

[31] BOGGS, P. T., BYRD, R. H., DONALDSON, J. R., AND SCHNABEL, R. B. *ORDPACK–software for weighted orthogonal distance regression.* Tech. rep CU-CS-360-87, Dept. of Comp. Sc., Univ. of Colorado, Boulder, CO, 1987.

[32] BOX, M., DAVIES, D., AND SWANN, W. *Nonlinear Optimization Techniques*, ICI Ltd. Monograph 5, Oliver and Boyd, Edinburgh, UK, 1969.

[33] BRODLIE, K. W. An assessment of two approaches to variable metric methods. *Math. Programming* 12 (1977), 344–355.

[34] BRODZIK, M. L. *Numerical approximation of manifolds and applications.* Ph.D. thesis, Dept. of Mathematics, Univ. of Pittsburgh, Pittsburgh, PA, 1996.

[35] BRODZIK, M. L., AND RHEINBOLDT, W. C. On the computation of simplicial approximations of implicitly defined two-dimensional manifolds. *Comput. Math. Appl.* 28 (1994), 9–21.

[36] BROOKS, F. Discussion of random methods for locating surface minima. *Oper. Res.* 6 (1958), 244–251.

[37] BROWDER, F. The solvability of nonlinear functional equations. *Duke Math. J.* 33 (1963), 557–567.

[38] BROWN, P. N., AND SAAD, Y. Hybrid Krylov methods for nonlinear systems of equations. *SIAM J. Sci. Stat. Comput.* 11 (1990), 450–481.

[39] BROYDEN, C. G. A class of methods for solving nonlinear simultaneous equations. *Math. Comp.* 19 (1965), 577–593.

[40] BROYDEN, C. G. Quasi-Newton methods and their application to function minimization. *Math. Comp.* 21 (1967), 368–381.

[41] Broyden, C. G. The convergence of a class of double–rank minimization algorithms. *J. Inst. Math. Appl.* 6 (1970), 76–90.

[42] Broyden, C. G. The convergence of an algorithm for solving sparse nonlinear systems. *Math. Comp.* 25 (1971), 285–294.

[43] Broyden, C. G. *Quasi-Newton methods.* in Numerical Methods for Unconstrained Minimization, W. Murray, ed. Academic Press, New York, 1972, pp. 87–106.

[44] Broyden, C. G., Dennis Jr, J. E., and Moré, J. J. On the local and superlinear convergence of quasi-Newton methods. *IMA J. Appl. Math.* 12 (1973), 223–246.

[45] Bryan, C. On the convergence of the method of nonlinear simultaneous displacements. *Rend. Circ. Mat. Palermo* 13 (1964), 177–191.

[46] Burkardt, J., and Rheinboldt, W. C. Algorithm 596: A program for a locally parametrized continuation process. *ACM Trans. Math. Software*, 9 (1983), 236–241.

[47] Burkardt, J., and Rheinboldt, W. C. A locally parametrized continuation process. *ACM Trans. Math. Software*, 9 (1983), 215–235.

[48] Cauchy, A. Méthode générale pour la résolution des systémes d'équations simultanes. *CR Acad. Sci. Paris* 25 (1847), 536–538.

[49] Cavanagh, R. *Difference equations and iterative methods,* Ph.D. thesis, Dept. of Mathematics, Univ. of Maryland, College Park, MD, 1970.

[50] Ciarlet, P. G. *The Finite Element Method for Elliptic Problems.* North–Holland, Amsterdam, New York, 1978.

[51] Coffmann, C. Asymptotic behavior of solutions of ordinary differential equations. *Trans. Amer. Math. Soc.* 110 (1964), 22–51.

[52] Coleman, T. F., Garbow, B. S., and Moré, J. J. Software for estimating sparse Jacobian marices: Algorithm 618. *ACM Trans. Math. Software*, 10 (1984), 329–347.

[53] Coleman, T. F., and Moré, J. J. Estimation of sparse Jacobian matrices and graph-coloring problems. *SIAM J. Numer. Anal.* 20 (1983), 187–209.

[54] Coleman, T. F., and Pothen, A. *The null-space problem II: Algorithms,* Tech. Rep. TR 86-747, Cornell Univ., Dept. of Computer Science, Ithaca, NY, 1986.

[55] Collatz, L. Aufgaben monotoner Art. *Arch. Math.* 3 (1952), 365–376.

[56] Curry, H. The method of steepest descent for nonlinear minimization problems. *Quart. Appl. Math.* 2 (1944), 258–261.

[57] Curry, J., Garnett, L., and Sullivan, D. On the iteration of rational functions: Computer experiments with Newton's method. *Comm. Math. Phys.* 91 (1983), 261–277.

[58] Curtis, A. R., Powell, M. J. D., and Reid, J. K. On the estimation of sparse Jacobian matrices. *J. Inst. Math. Appl.* 13 (1974), 117–119.

[59] Davidenko, D. On a new method of numerically integrating a system of nonlinear equations. *Dokl. Akad. Nauk SSSR* 88 (1953), 601–604. (In Russian.)

[60] Davidenko, D. On the approximate solution of a system of nonlinear equations. *Ukrain. Mat. Zh.* 5 (1953), 196–206 (in Russian).

[61] Davidon, W. *Variable metric methods for minimization,* Tech. Rep. ANL-5990, Argonne National Laboratory, Argonne, IL, 1959.

[62] Dembo, R., Eistenstat, S. C., and Steihaug, T. Inexact Newton methods. *SIAM J. Numer. Anal.* 19 (1982), 400–408.

[63] den Heijer, C., and Rheinboldt, W. C. On steplength algorithms for a class of continuation methods. *SIAM J. Numer. Anal.* 18 (1981), 925–948.

[64] Dennis Jr, J. E. *On some methods based on Broyden's secant approximation of the Hessian.* in Numerical Methods for Nonlinear Optimization, F. Lootsma, ed. Academic Press, New York, 1972.

[65] Dennis Jr, J. E., and Schnabel, R. B. Least change secant updates for Quasi–Newton methods. *SIAM Rev.* 21 (1979), 443–459.

[66] DENNIS JR, J. E., AND SCHNABEL, R. B. *Numerical Methods for Unconstrained Optimization and Nonlinear Equations.* Prentice–Hall, Englewood Cliffs, NJ, 1983.

[67] DENNIS JR, J. E., AND TURNER, K. Generalized conjugate directions. *J. Linear Algebra and Appl.* 88/89 (1986), 187–209.

[68] DENNIS JR, J. E., AND WALKER, H. F. Convergence theorems for least change secant update methods. *SIAM J. Numer. Anal.* 18 (1981), 949–987.

[69] DENNIS JR, J. E., AND WALKER, H. F. Least change sparse secant update methods with inaccurate secant condition. *SIAM J. Numer. Anal.* 22 (1985), 760–778.

[70] DEUFLHARD, P. A modified Newton method for the solution of ill–conditioned systems of nonlinear equations with application to multiple shooting. *Num. Math.* 22 (1974), 289–315.

[71] DEUFLHARD, P. A step size control for continuation methods and its special application to multiple shooting problems. *Numer. Math.* 33 (1979), 115–146.

[72] DEUFLHARD, P., FIEDLER, B., AND KUNKEL, P. Efficient numerical pathfollowing beyond critical points. *SIAM J. Numer. Anal.* 24 (1987), 912–927.

[73] DEUFLHARD, P., AND HEINDL, G. Affine invariant convergence theorems for Newton's method and extensions to related methods. *SIAM J. Numer. Anal.* 16 (1979), 1–10.

[74] DEUFLHARD, P., AND POTRA, F. Asymptotic mesh independence for Newton–Galerkin methods via a refined Mysovskii theorem. *SIAM J. Numer. Anal.* 29 (1992), 1395–1412.

[75] DEUFLHARD, P., AND WEISER, M. *Local inexact Newton multilevel FEM for nonlinear elliptic problems*, Tech. Rep. SC-96-29. K. Zuse Zentr. f. Informationstechnik, 1996.

[76] DIAZ, J., AND METCALF, F. On the set of subsequential limit points of successive approximations. *Trans. Amer. Math. Soc.* 135 (1969), 1–27.

[77] DIEUDONNE, J. *Foundations of Modern Analysis.* Academic Press, New York, 1960.

[78] DÖRING, B. Das Tschebyscheff Verfahren in Banach Räumen. *Numer. Math.* 15 (1970), 175–195.

[79] DÖRING, B. Einige Sätze über das Verfahren der tangentierenden Hyperbeln in Banach Räumen. *Appl. Math.* 15 (1970), 418–464.

[80] DÖRING, B. Über einige Klassen von Iterationsverfahren in Banach Räumen. *Math. Ann.* 187 (1970), 279–294.

[81] DOUGLAS JR., J. Alternating direction iteration for mildly nonlinear elliptic difference equations, I. *Numer. Math.* 3 (1961), 92–98.

[82] DOUGLAS JR., J. Alternating direction iteration for mildly nonlinear elliptic difference equations, II. *Numer. Math.* 4 (1962), 301–302.

[83] DOUGLAS JR., J. Alternating direction methods for three space variables. *Numer. Math.* 4 (1962), 41–63.

[84] DUFFIN, R. Nonlinear networks. *Bull. Amer. Math. Soc.* 54 (1948), 119–127.

[85] EAVES, C. Homotopies for the computation of fixed points. *Math. Programming* 3 (1972), 1–22.

[86] EAVES, C., AND SAIGAL, R. Homotopies for computation of fixed points on unbounded regions. *Math. Programming* 3 (1972), 225–237.

[87] EHRMANN, H. Konstruktion und Durchführung von Iterationsverfahren höherer Ordnung. *Arch. Rational Mech. Anal.* 4 (1959), 65–88.

[88] EISENSTAT, S. C., AND WALKER, H. F. *Choosing the forcing terms in an inexact Newton method*, Tech. Rep. 6/94/75, Dept. of Mathematics and Statistics, Utah State Univ., Logan, UT, 1994.

[89] EISENSTAT, S. C., AND WALKER, H. F. Globally convergent inexact Newton methods. *SIAM J. Optim.* 4 (1994), 393–422.

[90] ELKIN, R. *Convergence theorems for Gauss–Seidel and other minimization algorithms*, Ph.D. thesis, Dept. of Mathematics, Univ. of Maryland, 1968.

[91] FAN, K. Topological proofs for certain matrices with nonnegative elements. *Monatsh. Math.* 62 (1958), 219–237.

[92] FATOU, P. Sur les Equations Fonctionelles. *Bull. Soc. Math. France* 47 (1919), 161–271.

[93] FIEDLER, M., AND PTÁK, V. On matrices with non–positive off–diagonal elements and positive principal minors. *Czechoslovak Math. J.* 12 (1962), 382–400.

[94] FIEDLER, M., AND PTÁK, V. Some generalizations of positive definiteness and monotonicity. *Numer. Math.* 9 (1966), 163–172.

[95] FLETCHER, R. A new approach to variable metric algorithms. *Comput. J.* 13 (1970), 317–322.

[96] FLETCHER, R. *Practical methods of optimization, Vol. I, Unconstrained optimization.* John Wiley, New York, 1980.

[97] FLETCHER, R., AND POWELL, M. A rapidly convergent descent method for minimization. *Comput. J. 6* (1963), 163–168.

[98] FLETCHER, R., AND POWELL, M. On the modification of $LDL^{\top}$ factorizations. *Math. Comput.* 28 (1974), 1067–1087.

[99] FLETCHER, R., AND REEVES, C. Function minimization by conjugate gradients. *Comput. J.* 7 (1964), 149–154.

[100] FLOUDAS, C. A., AND PARDALOS, P. M. *State of the Art in Global Optimization.* Kluwer Academic Publishers, Dordrecht, Netherlands, 1996.

[101] FORSYTHE, G. E., AND WASOW, W. R. *Finite Difference Methods for Partial Differential Equations.* John Wiley, New York, 1960.

[102] FREUDENTHAL, H. Simplizialzerlegungen von beschränkter Flachheit. *Ann. of Math.* 43 (1942), 580–582.

[103] GAIER, D. *Konstruktive Methoden der konformen Abbildung.* Springer-Verlag, Heidelberg, New York, 1964.

[104] GAY, D. M. Subroutines for unconstrained minimization using a model/trust region approach. *ACM Trans. Math. Software* 9 (1983), 503–524.

[105] GEORG, K. *A note on stepsize control for numerical curve following.* in Homotopy Methods and Global Convergence, B. C. Eaves, F. J. Gould, H.-O. Peitgen, and M. J. Todd, eds. Plenum Press, New York, 1983, pp. 145–154.

[106] GILL, P. E., GOLUB, G. H., MURRAY, W., AND SAUNDERS, M. A. Methods for modifying matrix factorizations. *Math. Comp.* 28 (1974), 505–535.

[107] GILL, P. E., AND MURRAY, W. Quasi-Newton methods for unconstrained optimization. *J. Inst. Math. Appl.* 9 (1972), 91–108.

[108] GILL, P. E., MURRAY, W., AND SAUNDERS, M. A. Methods for computing and modifying the LDU factors of a matrix. *Math. Comp.* 29 (1975), 1051–1077.

[109] GILL, P. E., MURRAY, W., AND WRIGHT, M. H. *Practical Optimization.* Academic Press, New York, 1981.

[110] GLOWINSKI, R. *Numerical Methods for Nonlinear Variational Problems.* Springer-Verlag, Heidelberg, New York, 1984.

[111] GOLDFARB, D. A family of variable metric methods derived by variational means. *Math. Comp.* 24 (1970), 23–26.

[112] GOLDFARB, D. Factorized variable metric methods for unconstrained optimization. *Math. Comp.* 30 (1976), 796–811.

[113] GOLDSTEIN, A. *Constructive Real Analysis.* Harper and Row, New York, 1967.

[114] GOLUB, G. H., AND VAN LOAN, C. F. *Matrix Computations*, 2nd ed. The Johns Hopkins University Press, Baltimore, MD, 1989.

[115] GREENSPAN, D. *Introductory Numerical Analysis of Elliptic Boundary Value Problems.* John Wiley, New York, 1965.

[116] GREENSPAN, D., AND PARTER, S. Mildly nonlinear elliptic partial differential equations and their numerical solution, II. *Numer. Math.* 7 (1965), 129–147.

[117] Greenstadt, J. Variations on variable-metric methods. *Math. Comp.* 24 (1970), 1–18.

[118] Greenspan, D., and Yohe, M. On the approximate solution of $\Delta u = F(u)$. *Comm. ACM* 6 (1963), 564–568.

[119] Griewank, A., and Corliss, G., eds. *Automatic Differentiation of Algorithms: Theory, Implementation and Application.* SIAM, Philadelphia, PA, 1992.

[120] Gunn, J. On the two-stage iterative method of Douglas for mildly nonlinear elliptic difference equations. *Numer. Math.* 6 (1964), 243–249.

[121] Gurin, L., and Rastrigin, L. Convergence of the random search method in the presence of noise. *Avtomat. i Telemekh.* 26 (1965), 1505–1511.

[122] Hachtel, G. D., Brayton, R. K., and Gustavson, F. G. The sparse tableau approach to network analysis and design. *IEEE Trans. Circuit Theory CT*-18 (1971), 101–113.

[123] Hackbusch, W. *Iterative Solution of Large Sparse Systems of Equations*, Appl. Math. Sci, 95, Springer-Verlag, Heidelberg, New York, 1993.

[124] Heitzinger, W., Troch, L., and Valentin, G. *Praxis nichtlinearer Gleichungen.* Carl Hanser-Verlag, München, Germany, 1985. (In German.)

[125] Henderson, M. E. *Computing implicitly defined surfaces: Two parameter continuation*, Tech. Rep. RC-18777 (82115), IBM T. J. Watson Research Center, Yorktown Heights, NY, 1993.

[126] Hestenes, M., and Stiefel, E. Methods of conjugate gradients for solving linear systems. *J. Res. Nat. Bur. Stand.* 49 (1952), 409–436.

[127] Heydt, G. *Directed random search*, Ph.D. thesis, Purdue Univ., Lafayette, IN, 1970.

[128] Hohmann, A. *An adaptive continuation method for implicitly defined surfaces*, Tech. Rep. SC 91-20, K. Zuse Zentrum f. Inf. Tech., Berlin, Germany, 1991.

[129] Hohmann, A. *Inexact Gauss–Newton methods for parameter dependent nonlinear problems*, Ph.D. thesis, Inst. f. Mathematik, F. U. Berlin, Berlin, Germany, 1994.

[130] Horst, R., and Pardalos, P. M. *Handbook of Global Optimization.* Kluwer Academic Publishers, Dordrecht, Netherlands, 1995.

[131] Horst, R., Pardalos, P. M., and Thoai, N. V. *Introduction to Global Optimization.* Kluwer Academic Publishers, Dordrecht, Netherlands, 1996.

[132] Horst, R., and Tuy, H. *Global optimization*, 2nd ed. Springer-Verlag, Heidelberg, New York, 1993.

[133] Householder, A. S. *The Numerical Treatment of a Single Nonlinear Equation.* McGraw–Hill, New York, 1970.

[134] Householder, A. S. *The Theory of Matrices in Numerical Analysis.* Dover, New York, 1974.

[135] Huang, H. Unified approach to quadratically convergent algorithms for function minimization. *Optim. Theory and Appl.* 5 (1070), 405–423.

[136] Jacoby, S., Kowalik, J., and Pizzo, J. *Iterative Methods for Nonlinear Optimization Problems.* Prentice–Hall, Englewood Cliffs, NJ, 1972.

[137] Julia, G. Memoire sur l'iteration des fonctions rationelles. *J. Math. Pures Appl.* 4 (1918), 47–245.

[138] Kačurovskii, R. On monotone operators and convex functionals. *Uspekhi Mat. Nauk* 15 (1960), 213–215 (in Russian).

[139] Kahan, W. *Gauss–Seidel methods for solving large systems of linear equations*, Ph.D. thesis, Univ. of Toronto, Toronto, Canada, 1958.

[140] Kantorovich, L. V. The method of successive approximations for functional equations. *Acta Math.* 71 (1939), 63–97.

[141] Kearfott, R. B. *Rigorous Global Search: Continuous Problems.* Kluwer Academic Publishers, Dordrecht, Netherlands, 1996.

[142] Keller, H. B. *Numerical Methods for Two-point Boundary Value Problems.* Blaisdell, Waltham, MA, 1968.

[143] KELLER, H. B. *Numerical Methods for Two-point Boundary Value Problems*, Reg. Conf. Ser. Appl. Math, 24, SIAM, Philadelphia, PA, 1976.

[144] KELLER, H. B. *Numerical solution of bifurcation and nonlinear eigenvalue problems.* in Applications of Bifurcation Theory, P. H. Rabinowitz, ed. Academic Press, New York, 1977, pp. 359–384.

[145] KELLER, H. B. *Lectures on Numerical Methods in Bifurcation Problems.* Springer-Verlag, Heidelberg, New York, 1987.

[146] KELLEY, C. T. *Iterative Methods for Linear and Nonlinear Equations*, Frontiers in Appl. Math, 16, SIAM, Philadelphia, PA, 1995.

[147] KELLOGG, R. B. A nonlinear alternating direction method. *Math. Comp.* 23 (1969), 23–28.

[148] KERNER, M. Die Differentiale in der allgemeinen Analysis. *Ann. Math.* 34 (1933), 546–572.

[149] KNUTH, D. E. *The Art of Computer Programming, Vol. 2: Semi-numerical Algorithms.* Addison–Wesley, Reading, MA, 1971.

[150] KUHN, H. W. Some combinatorial lemmas in topology. *IBM J. Res. Devel.* 4/5 (1960), 518–524.

[151] KUHN, H. W. *Approximate search for fixed points.* in Computing Methods in Optimization Problems 2, L. A. Zadek, L. W. Neustadt, and A. V. Balakrishnan, eds. Academic Press, New York, 1969, pp. 199–211.

[152] LAWRENCE, J., AND STEIGLITZ, K. Randomized pattern search. *IEEE Trans. Comput.* *C*-21 (1972), 382–385.

[153] LEMKE, C. E. Bi-matrix equilibrium points and mathematical programming. *Manag. Sci.* 11 (1965), 681–689.

[154] LEMKE, C. E., AND HOWSON, J. T. Equilibrium points of bi-matrix games. *SIAM J. Appl. Math.* 12 (1964), 413–423.

[155] LEVENBERG, K. A method for the solution of certain nonlinear problems in least squares. *Quart. Appl. Math.* 2 (1944), 164–168.

[156] LIEBERSTEIN, H. *Overrelaxation for nonlinear elliptic partial differential equations*, Tech. Rep. 80, Mathematics Resource Center, Univ. of Wisconsin, Madison, WI, 1959.

[157] LUCHKA, A. *Theory and Application of the Averaging Method for Functional Corrections.* Academic Press, New York, 1965. (In Russian.)

[158] LUDWIG, R. Verbessserung einer Iterationsfolge bei Gleichungssystemen. *Z. Angew. Math. Mech.* 32 (1952), 232–234.

[159] LUNDBERG, B. N., AND POORE, A. B. Variable order Adams–Bashforth predictors with error-stepsize control for continuation methods. *SIAM J. Sci. Stat. Comput.* 12 (1991), 695–723.

[160] MACKENS, W. Numerical differentiation of implicitly defined space curves. *Computing* 41 (1989), 237–260.

[161] MARCHUK, G. I. *Methods of Numerical Mathematics*, 2nd ed. Springer-Verlag, Heidelberg, New York, 1992.

[162] MARCHUK, G. I., AND SHAIDUROV, V. V. *Difference Methods and Their Extrapolations.* Springer-Verlag, Heidelberg, New York, 1983.

[163] MARQUARDT, D. An algorithm for least squares estimation of nonlinear parameters. *SIAM J. Appl. Math.* 11 (1963), 431–441.

[164] MATYAS, J. Random optimization. *Avtomat. i Telemekh.* 26 (1965), 246–253.

[165] MCCORMICK, G. P. A modification of Armijo's step-size rule for negative curvature. *Math. Programming* 13 (1977), 111–115.

[166] MELVILLE, R., AND MACKEY, S. A new algorithm for two-dimensional continuation. *Comput. Math. Appl.* 30 (1995), 31–46.

[167] MERRILL, O. *Applications and extensions of an algorithm that computes fixed points of a certain upper semi-continuous point to set mapping*, Ph.D. thesis, Dept. of Industrial Engineering, Univ. of Michigan, Ann Arbor, MI, 1972.

[168] MERTVECOVA, M. An analogue of the method of osculating hyperbolas for general operator equations. *Dokl. Akad. Nauk SSSR* 88 (1953), 611–614.

[169] MEYER, G. *The numerical solution of quasilinear equations.* in Numerical Solution of Systems of Nonlinear Algebraic Equations, G. Byrne and C. A. Hall, eds. Academic Press, New York, 1973, pp. 27–62.

[170] MILNOR, J. W. *Topology from the Differentiable Viewpoint.* University Press of Virginia, Charlottesville, VA, 1969.

[171] MINTY, G. Monotone (nonlinear) operators in Hilbert space. *Duke Math. J.* 29 (1962), 341–346.

[172] MITCHELL, A. R., AND GRIFFITHS, D. F. *The Finite Difference Method in Partial Differential Equations.* John Wiley, New York, 1980.

[173] MITTELMANN, H. D., AND ROOSE, D., eds. *Continuation techniques and bifurcation problems*, Int. Ser. Numer. Math., 92, Birkhäuser-Verlag, Basel, Switzerland, 1990.

[174] MORÉ, J., AND RHEINBOLDT, W. C. On P- and S-functions and related classes of nonlinear mappings. *Linear Algebra Appl.* 6 (1973), 45–68.

[175] MORÉ, J. J. Global convergence of Newton–Gauss–Seidel methods. *SIAM J. Numer. Anal.* 8 (1971), 325–336.

[176] MORÉ, J. J. Nonlinear generalizations of matrix diagonal dominance with applications to Gauss–Seidel iterations. *SIAM J. Numer. Anal.* 9 (1972), 357–378.

[177] MORÉ, J. J. Classes of functions and feasibility conditions in nonlinear complementarity problems. *Math. Programming* 6 (1974), 327–338.

[178] MORÉ, J. J. Coercivity conditions in nonlinear complementarity. *SIAM Rev.* 16 (1974), 1–16.

[179] MORÉ, J. J. *A collection of nonlinear model problems.* in Computational Solution of Nonlinear Systems of Equations, E. L. Allgower and K. Georg, eds. American Mathematical Society, Providence, RI, 1990, pp. 723–762.

[180] MORÉ, J. J., AND SORENSEN, D. C. On the use of directions of negative curvature in a modified Newton method. *Math. Programming* 16 (1979), 1–20.

[181] MORÉ, J. J., AND SORENSEN, D. C. Computing a trust region step. *SIAM J. Sci. Stat. Comput.* 4 (1983), 553–572.

[182] MORGAN, A. P. *Solving Polynomial Systems Using Continuation for Engineering and Scientific Problems.* Prentice–Hall, Englewood Cliffs, NJ, 1987.

[183] NAZARETH, J. L. The method of successive affine reduction for nonlinear minimization. *Math. Programming* 35 (1986), 97–109.

[184] NECEPURENKO, M. On the Chebyshev method for operator equations. *Uspekhi Mat. Nauk* 9 (1954), 163–170.

[185] NELDER, J., AND MEAD, R. A simplex method for function minimization. *Comput. J.* 7 (1965), 308–313.

[186] NICKEL, K., AND RITTER, K. Termination criterion and numerical convergence. SIAM J. Numer. Anal. 9 (1972), 277–283.

[187] NIKAIDO, H. *Convex Structures and Economic Theory.* Academic Press, New York, 1968.

[188] NUSSE, H. E., AND YORKE, J. A. *Dynamics: Numerical Explorations*, Appl. Math. Sci., Springer-Verlag, Heidelberg, New York, 1994.

[189] NUSSE, H. E., AND YORKE, J. A. Basins of attraction. *Science* 271 (March 1996), 1376–1380.

[190] ORTEGA, J., AND RHEINBOLDT, W. C. A general convergence result for unconstrained minimization methods. *SIAM J. Numer. Anal.* 9 (1972), 40–43.

[191] ORTEGA, J. M., AND RHEINBOLDT, W. C. Monotone iterations for nonlinear equations with applications to Gauss–Seidel methods. *SIAM J. Numer. Anal.* 4 (1967), 171–190.

[192] ORTEGA, J. M., AND RHEINBOLDT, W. C. *Iterative Solutions of Nonlinear Equations in Several Variables.* Academic Press, New York, 1970.

[193] ORTEGA, J. M., AND RHEINBOLDT, W. C. *Local and global convergence of generalized linear iterative methods.* in Numerical Solution of Nonlinear Problems, J. M. Ortega and W. C. Rheinboldt, eds., SIAM, Philadelphia, PA, 1970, pp. 122–143.

[194] ORTEGA, J. M., AND ROCKOFF, M. Nonlinear difference equations and Gauss–Seidel type iterative methods. *SIAM J. Numer. Anal.* 3 (1966), 497–513.

[195] OSTROWSKI, A. M. Über die Determinanten mit überwiegender Haupdiagonale. *Comm. Math. Helv.* 10 (1937), 69–96.

[196] OSTROWSKI, A. M. *Solution of Equations and Systems of Equations.* Academic Press, New York, 1960.

[197] OSTROWSKI, A. M. *Solution of Equations in Euclidean and Banach Spaces*, 3rd ed. Academic Press, New York, 1973.

[198] PARKINSON, J., AND HUTCHINSON, D. *An investigation into the effectiveness of variants on the simplex method.* in Numerical Methods for Nonlinear Optimization, F. Lootsma, ed. Academic Press, New York, 1972, pp. 115–136.

[199] PARTHASARATHY, T. *On Global Univalence Theorems*, Lecture Notes in Math., Springer-Verlag, Heidelberg, New York, 1983.

[200] PEARSON, J. Variable metric methods for solving equations and inequalities. *Comput. J.* 12 (1969), 171–178.

[201] PEITGEN, H.-O., AND RICHTER, P. H. *The Beauty of Fractals.* Springer-Verlag, Heidelberg, New York, 1986.

[202] PERRON, O. Über Stabilität und asymptotisches Verhalten der Lösungen eines Systems endlicher Differenzengleichungen. *J. Reine Angew. Math.* 161 (1929), 41–64.

[203] PORSCHING, T. A. Jacobi and Gauss–Seidel methods for nonlinear network problems. *SIAM J. Numer. Anal.* 6 (1969), 437–449.

[204] PORSCHING, T. A. On rates of convergence of Jacobi and Gauss-Seidel for M–functions. *SIAM J. Numer. Anal.* 8 (1971), 575–582.

[205] POWELL, M. Recent advances in unconstrained optimization. *Math. Programming* 1 (1971), 26–57.

[206] POWELL, M. J. D. *A new algorithm for unconstrained optimization.* in Nonlinear Programming, J. B. Rosen, O. L. Mangasarian, and K. Ritter, eds. Academic Press, New York, 1970, pp. 31–65.

[207] RASTRIGIN, L. The convergence of the random search method in the extremal control of a many parameter system. *Avtomat. i Telemekh.* 24 (1963), 1467–1473.

[208] RATSCHEK, H., AND ROKNE, J. *New Computer Methods for Global Optimization.* Ellis–Norwood, Chichester, UK, 1988.

[209] RENEGAR, J. On the efficiency of Newton's method in approximating all zeros of a system of complex polynomials. *Math. Oper. Res.* 12 (1987), 121–148.

[210] RHEINBOLDT, W. C. On M-functions and their application to nonlinear Gauss-Seidel iterations and to network flows. *J. Math. Anal. Appl.* 32 (1970), 274–307.

[211] RHEINBOLDT, W. C. *On classes of n-dimensional nonlinear mappings generalizing several types of matrices.* in Numerical Solution of Partial Differential Equations II, B. Hubbard, ed. Academic Press, New York, 1971, pp. 501–545.

[212] RHEINBOLDT, W. C. Solution fields of nonlinear equations and continuation methods. *SIAM J. Numer. Anal.* 17 (1980), 221–237.

[213] RHEINBOLDT, W. C. *On a constructive moving frame algorithm and the triangulation of equilibrium manifolds.* in Bifurcation: Analysis Algorithms Applications, T. Kuepper, R. Seydel, and H. Troger, eds., Int. Ser. Numer. Math., 79, Birkhäuser-Verlag, Basel, Switzerland, 1987, pp. 256–267.

[214] RHEINBOLDT, W. C. On the computation of multi-dimensional solution manifolds of parametrized equations. *Numer. Math.* 53 (1988), 165–181.

[215] RHEINBOLDT, W. C. MANPACK: A set of algorithms for computations on implicitly defined manifolds. *Comput. Math. Appl.* 27 (1996), 15–28.

[216] ROBINSON, S. Interpolative solution of systems of nonlinear equations. *SIAM J. Numer. Anal.* 3 (1966), 650–658.

[217] ROSENBROCK, H. An automatic method for finding the greatest or least value of a function. *Comput. J.* 3 (1960), 175–184.

[218] SAAD, Y., AND SCHULTZ, M. H. GMRES: a generalized minimal residual method for solving nonsymmetric linear systems. *SIAM J. Sci. Stat. Comput.* 7 (1986), 856–869.

[219] SAFIEV, R. The method of tangent hyperbolas. *Soviet Math. Dokl.* 4 (1963), 482–485.

[220] SALANE, D. E. A continuation approach for solving large–residual nonlinear least squares problems. *SIAM J. Sci. Stat. Comput.* 8 (1987), 655–671.

[221] SAMANSKII, V. On a modification of the Newton method. *Ukrain. Mat. Zh.* 19 (1967), 788 790.

[222] SANDBERG, I., AND WILLSON, A. Some network theoretic properties of nonlinear DC transistor networks. *Bell Syst. Tech. J.* 48 (1969), 1–34.

[223] SANDBERG, I., AND WILLSON, A. Some theorems on properties of DC equations of nonlinear networks. *Bell Syst. Tech. J.* 48 (1969), 1293–1311.

[224] SCARF, H. E. The approximation of fixed points of a continuous mapping. *SIAM J. Appl. Math.* 15 (1967), 1328–1343.

[225] SCARF, H. E., AND HANSEN, T. *The Computation of Economic Equilibria*, Cowles Found. Monograph, 24, Yale University Press, New Haven, CT, 1973.

[226] SCHECHTER, S. Iteration methods for nonlinear problems. *Trans. Amer. Math. Soc.* 104 (1962), 179–189.

[227] SCHECHTER, S. Relaxation methods for convex problems. *SIAM J. Numer. Anal.* 5 (1968), 601–612.

[228] SCHECHTER, S. *On the choice of relaxation parameters for nonlinear problems.* in Numerical Solution of Systems of Nonlinear Algebr. Equations, G. D. Byrne and C. A. Hall, eds. Academic Press, New York, 1973.

[229] SCHMIDT, J. Die Regula Falsi für Operatoren in Banachräumen. *Z. Angew. Math. Mech.* 41 (1961), 61–63.

[230] SCHMIDT, J., AND TRINKAUS, H. Extremwertermittlung mit Funktionswerten bei Funktionen von mehreren Veränderlichen. *Computing* 1 (1966), 224–232.

[231] SCHNABEL, R. B., KOONTZ, J. E., AND WEISS, B. E. A modular system of algorithms of unconstrained minimization. *ACM Trans. Math. Software* 11 (1985), 419–440.

[232] SCHRACK, G., AND BOROWSKI, N. *An experimental comparison of three random searches.* in Numerical Methods in Nonlinear Optimization, F. Lootsma, ed. Academic Press, New York, 1972, pp. 137–148.

[233] SCHRÖDER, E. Über unendlich viele Algorithmen zur Auflösung der Gleichungen. *Math. Ann.* 2 (1870), 317–365.

[234] SCHRÖDER, J. Range-domain implications for linear operators. *SIAM J. Appl. Math.* 19 (1970), 235–242.

[235] SCHUBERT, L. Modification of a quasi-Newton method for nonlinear equations with a sparse Jacobian. *Math. Comp.* 24 (1970), 27–30.

[236] SCHUMER, M., AND STEIGLITZ, K. Adaptive step–size random search. *IEEE Trans. Automat. Control* AC-13 (1968), 270–276.

[237] SCHWARZ, H. R. *Methode der finiten Elemente*, 2nd ed. Teubner, Stuttgart, Germany, 1984.

[238] SCHWETLICK, H., AND TILLER, V. Numerical methods for estimating parameters in nonlinear models with errors in the variables. *Technometrics* 27 (1985), 17–24.

[239] SEDGEWICK, R. *Algorithms.* Addison–Wesley, Reading, MA, 1984.

[240] SEIFERT, H., AND THRELFALL, W. *Lehrbuch der Topologie.* Chelsea Publ. Co, 1947.

[241] SESHU, S., AND REED, H. *Linear Graphs and Electrical Networks.* Addison–Wesley, Reading, MA, 1961.

[242] SEYDEL, R. *From Equilibrium to Chaos.* Elsevier, New York, 1988.

[243] SEYDEL, R., W.SCHNEIDER, F., KÜPPER, T., AND TROGER, H., eds. *Bifurcation and Chaos: Analysis, Algorithms, Applications,* Int. Ser. Numer. Math, 97, Birkhäuser-Verlag, Basel, Switzerland, 1991.

[244] SHANNO, D. F. Conditioning of quasi-Newton methods for function minimization. *Math. Comp.* 24 (1970), 647–656.

[245] SHANNO, D. F., AND PHUA, K. H. Numerical comparison of several variable metric algorithms. *J. Optim. Theory Appl.* 25 (1978), 507–518.

[246] SHUB, M., AND SMALE, S. Computational complexity: On the geometry of polynomials and a theory of cost. *SIAM J. Comput.* 15 (1986), 145–161.

[247] SLEPIAN, P. *Mathematical Foundations of Network Analysis,* Tracts in Nat. Philos., 16, Springer-Verlag, Heidelberg, New York, 1968.

[248] SMALE, S. On the mathematical foundations of electrical circuit theory. *J. Differential Geom.* 7 (1972), 193–210.

[249] SMALE, S. On the efficiency of algorithms of analysis. *Bull. Amer. Math. Soc.* 13 (1985), 87–121.

[250] SORENSEN, D. C. The q-superlinear convergence of a collinear scaling algorithm for unconstrained minimization. *SIAM Rev.* 17 (1980), 84–114.

[251] SORENSEN, D. C. Newton's method with a model trust–region modification. *SIAM J. Numer. Anal.* 19 (1982), 409–426.

[252] SPENDLEY, W., HEXT, G., AND HIMSWORTH, F. Sequential application of simplex designs in optimization and evolutionary operation. *Technometrics* 4 (1962), 441–461.

[253] SPERNER, E. Neuer Beweis für die Invarianz der Dimensionszahl. *Abh. Math. Sem. Univ. Hamburg* 6 (1928), 265–272.

[254] SPIVAK, M. *A Comprehensive Introduction to Differential Geometry, Vol. 1–5.* Publish or Perish, Berkeley, CA, 1979.

[255] STEPLEMAN, R. Finite dimensional analogues of variational problems in the plane. *SIAM J. Numer. Anal.* 8 (1971).

[256] STEWART, G. W. *Introduction to Matrix Computations.* Academic Press, New York, 1973.

[257] STRUWE, M. *Variational Methods.* Springer-Verlag, Heidelberg, New York, 1990.

[258] SZABO, B., AND BABUSKA, I. *Finite Element Analysis.* John Wiley, New York, 1991.

[259] THACKER, W. C. *Large least–squares problems and the need for automating the generation of adjoint code.* in Computational Solution of Nonlinear Systems of Equations, E. L. Allgower and K. Georg, eds. American Mathematical Society, Providence, RI, 1990, pp. 645–677.

[260] TODD, M. J. *The computation of fixed points and applications,* Lecture Notes in Econ. and Math. Syst., 124, Springer-Verlag, Heidelberg, New York, 1976.

[261] TOINT, P. L. On the superlinear convergence of an algorithm for solving a sparse minimization problem. *SIAM Rev.* 16 (1979), 1036–1045.

[262] TOINT, P. L. On large scale nonlinear least squares calculations. *SIAM J. Sci. Stat. Comput.* 8 (1987), 416–435.

[263] TÖRN, A., AND ZILINSKAS, A. *Global Optimization.* Springer-Verlag, Heidelberg, New York, 1989.

[264] TÖRNIG, W., GIPSER, M., AND KASPAR, B. *Numerische Lösung von partiellen Differentialgleichungen der Technik. Differenzenverfahren, finite Elemente und die Behandlung grosser Gleichungssysteme.* Teubner, Stuttgart, Germany, 1985. (In German.)

[265] TRAUB, J. F. *Iterative Methods for the Solution of Equations.* Prentice–Hall, Englewood Cliffs, NJ, 1964.

[266] TRAUB, J. F., WASILKOWSKI, G. W., AND WOŹNIAKOWSKI, H. *Information Based Complexity.* Academic Press, New York, 1988.

[267] TRAUB, J. F., AND WOŹNIAKOWSKI, H. *A General Theory of Optimal Algorithms.* Academic Press, New York, 1980.

[268] TURNER, K., AND WALKER, H. F. Efficient high accurracy solutions with GMRES(m). *SIAM J. Sci. Stat. Comput.* 13 (1992), 815–825.

[269] VARGA, R. S. *Matrix Iterative Analysis.* Prentice–Hall, Englewood Cliffs, NJ, 1962.

[270] VOIGT, R. Rates of convergence of iterative methods for nonlinear systems of equations. *SIAM J. Numer. Anal.* 8 (1971), 127–134.

[271] VRSCAY, E. R. Julia sets and Mandelbrot–like sets associated with higher order Schroeder iteration functions. *Math. Comput.* 46 (1986), 151–169.

[272] WALKER, H. F. Implementation of the GMRES method using Householder transformations. *SIAM J. Sci. Stat. Comput.* 9 (1988), 152–163.

[273] WALKER, H. F. Implementations of the GMRES method. *Comput. Phys. Comm.* 53 (1989), 311–320.

[274] WALKER, H. F. *A GMRES–backtracking Newton iterative method*, Tech. Rep. 3/94/74, Dept. of Mathematics and Statistics, Utah State Univ., Logan, UT, 1994.

[275] WATSON, L. T., BILLUPS, S. C., AND MORGAN, A. P. Algorithm 652: HOMPACK: A suite of codes for globally convergent homotopy algorithms. *ACM Trans. Math. Software* 13 (1987), 281–310.

[276] WATSON, L. T., SOSONKINA, M., MELVILLE, R. C., MORGAN, A. P., AND WALKER, H. P. *HOMPACK90: A suite of FORTRAN 90 codes for globally convergent homotopy algorithms*, Tech. Rep. TR 96-11, Dept. of Comput. Science, Virginia Polytech. Inst. and State Univ., Blacksburg, VA, 1996.

[277] WEGGE, J. On a discrete version of the Newton-Raphson method. *SIAM J. Numer. Anal.* 3 (1966), 134–142.

[278] WHITNEY, H. Analytic extensions of differentiable functions defined in closed sets. *Trans. Amer. Math. Soc.* 36 (1934), 63–89.

[279] WHITNEY, H. *Geometric Integration Theory.* Princeton University Press., Princeton, NJ, 1957.

[280] WITTICH, H. Bemerkungen zur Druckverteilungsrechnung nach Theodorsen–Garrick. *Jahrb. dtsch. Luftf.-Forsch.* (1941), I 52–I 57.

[281] WOLFE, P. The secant method for simultaneous nonlinear equations. *Comm. ACM.* 2 (1959), 12–13.

[282] WOLFE, P. Convergence conditions for ascent methods, I. *SIAM Rev.* 11 (1969), 226–235.

[283] WOLFE, P. Convergence conditions for ascent methods, II. *SIAM Rev.* 13 (1971), 185–188.

[284] YOUNG, D. M. *Iterative Solution of Large Linear Systems.* Academic Press, New York, 1971.

[285] YPMA, T. J. Efficient estimation of sparse Jacobian matrices by differences. *J. Comput. Appl. Math.* 18 (1987), 17–28.

[286] ZIENKIEWICZ, O. C. *The Finite Element Method*, 3rd ed. McGraw Hill Book Co, New York, London, 1977.

Index

UNIVERSITY
OF BRISTOL
LIBRARY
ENGINEERING